KU-289-982

014045643

The Engineer's Contribution to Contemporary Architecture

PETER RICE

WITHDRAWN FROM STOCK

Series editors

Angus Macdonald
Remo Pedreschi

Department of Architecture
University of Edinburgh

The Engineer's Contribution to Contemporary Architecture

PETER RICE

André Brown

Thomas Telford

Endorsed by

RIBA Publications

Published by Thomas Telford Publishing, Thomas Telford Ltd, 1 Heron Quay, London E14 4JD.
URL: http://www.thomastelford.com

Distributors for Thomas Telford books are
USA: ASCE Press, 1801 Alexander Bell Drive, Reston, VA 20191-4400, USA
Japan: Maruzen Co. Ltd, Book Department, 3-10 Nihonbashi 2-chome, Chuo-ku, Tokyo 103
Australia: DA Books and Journals, 648 Whitehorse Road, Mitcham 3132, Victoria

First published 2001

Also available from Thomas Telford Books

Engineer's Contribution to Contemporary Architecture - Eladio Dieste. R. Pedreschi ISBN 0 7277 2772 9
Engineer's Contribution to Contemporary Architecture - Anthony Hunt. A. Macdonald ISBN 0 7277 2769 9
Engineer's Contribution to Contemporary Architecture - Heinz Isler. J. Chilton ISBN 0 7277 2878 4
Engineer's Contribution to Contemporary Architecture - Owen Williams. D. Yeomans and D. Cottam ISBN 0 7277 3018 5

A catalogue record for this book is available from the British Library

ISBN: 0 7277 2770 2

© André Brown and Thomas Telford Limited 2001

All rights, including translation, reserved. Except as permitted by the Copyright, Designs and Patents Act 1988, no part of this publication may be reproduced, stored in a retrieval system or transmitted in any form or by any means, electronic, mechanical, photocopying or otherwise, without the prior written permission of the Publishing Director, Thomas Telford Publishing, Thomas Telford Ltd, 1 Heron Quay, London E14 4JD.

This book is published on the understanding that the author is solely responsible for the statements made and opinions expressed in it and that its publication does not necessarily imply that such statements and/or opinions are or reflect the views or opinions of the publishers. While every effort has been made to ensure that the statements made and the opinions expressed in this publication provide a safe and accurate guide, no liability or responsibility can be accepted in this respect by the author or publishers.

Designed by Acrobat
Printed and bound in Great Britain by the Cromwell Press, Trowbridge, Wiltshire

Acknowledgements

It is impossible to write a book like this without the co-operation of and help from a range of people who knew the subject and the subject's work. The problem with someone so multicultural and eclectic as Peter Rice is that it is impossible to speak to all of those who might have a view. So, first, apologies to those whom I was not able to consult.

There are a number of people, however, whom I do need to thank.

First of all my sincere thanks must go to Sylvia Rice who gave me access to Peter Rice's private notes and sketches. They were invaluable in gaining an understanding of the way that a range of issues were understood and worked through.

On the engineering and design side of the Arup organisation I am indebted to Bob Emmerson, John Nutt, Kenneth Fraser and Tristram Carfrae who spared the time to talk about someone who had clearly been both a friend and a colleague. Jane Wernick, who worked with Peter Rice at Arups, gave me a valuable insight into how he worked with others in the team. In the Arup Library Pauline Shirley and her assistant, Petra Lindahl, always gave willing cooperation.

In the Richard Rogers Partnership Vicki McGregor was very helpful. And at RFR in Paris my thanks go to Bernard Vaudeville, Dan Burr, Keiron Rice and Matthieu David. *Merci* to all of you.

Finally, thanks to Adam and Hannah for staying off the computer and for not messing up the slides.

André Brown

June 2001

Preface

Rather than write a preface to this book it is better that I let others set the scene and write most of it for me. First, Peter Rice himself in his RIBA Royal Gold Medal Speech.[1]

'If I have a philosophy, if I have a belief, it is the contribution that we can make and the contribution that we should make, is not to be quasi-architects. People often call me an architect–engineer, that's a lot of rubbish, I am an engineer, plain and simple.'

For the 21 years from the time of the Beaubourg project, where their collaboration and friendships really began, Renzo Piano and Richard Rogers maintained personal and professional relationships with Peter Rice. Relationships that were born out of mutual trust and respect. These are some of their words about their valued colleague.

'Like his great predecessors, whether Brunel or Brunelleschi, Peter Rice is able to step outside the confines of his professional training, transferring technical problems into poetical solutions. His design combines order with delight, science with art.' Richard Rogers[2]

'Peter was a man of science, and a great humanist, in the renaissance sense of the word. He never accepted the banal solution or the quick fix; it was from him that I learned never to be satisfied. Peter always said: 'Genius is great patience'. Renzo Piano[3]

Next Rice's architect collaborator in the RFR practice, Ian Ritchie.

'... the more his confidence grew... the more he felt he was simultaneously engineer, licensed dreamer and poet... [and we shared]... a need for contemporary architecture to communicate more than just a 'coup-visuel', and to find sensual expression that the ordinary person could feel.'

And finally from Jack Zunz, on many occasions Rice's guide and mentor at Arups.[4]

'His optimistic presence in a pessimistic world leaves a void which is difficult to fill.'

This book is about explaining the considerable extent of that void.

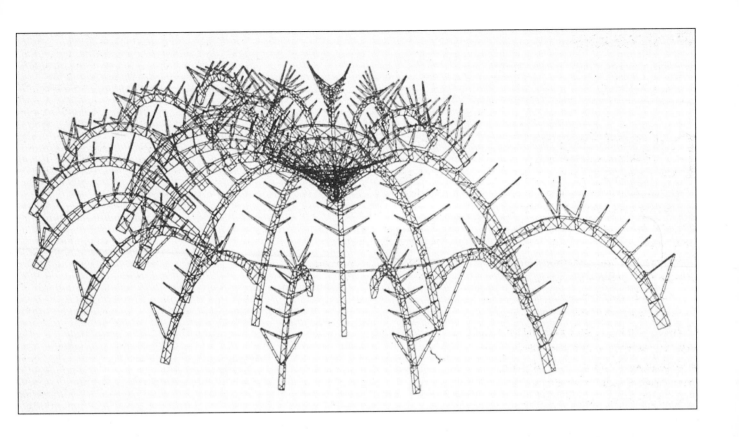

Contents

Chapter One
Rice, engineering and architecture

Photo: Ove Arup and Partners

Fig. 1.1. Peter Rice.

> *Peter Rice is one of those engineers who has greatly contributed to architecture, re-affirming the deep creative interconnection between humanism and science, between art and technology.*
>
> Renzo Piano[5]

Confident, competent and a risk taker

Peter Rice was born in Dundalk, Ireland in 1935 into an environment that he openly admitted was devoid of positive engineering or architectural influences. If you were to paint an architectural picture of the buildings and landscape that he described as surrounding him the palette would be drawn from a narrow spectrum from pale to mid grey. Nor was there a particular family or social influence either to explain the extraordinarily gifted talent that was to emerge from this unlikely starting point.

He studied engineering, apparently by accident, and accidentally found himself, in 1956, at Arups, clutching his engineering degree. After a year at Arups he went to Imperial College to undertake postgraduate studies, returning to Arups in 1958.

He said that he actually imagined when he was younger, that he would be a filmmaker, but somehow engineering crept up on him. In others such a story might be taken to be an embellishment to support the notion of an innate artistic talent that was to blossom later: but not in Peter Rice. In his case this is simply an account of what happened, presented in a typically self-effacing and direct way.

When the fame and adulation came later that was something that he really revelled in. But the journey to getting there seems, genuinely, to have been one that

began in an unexpected direction from an unlikely starting point.

Rice's first major project was the Sydney Opera House, where he was initially a junior engineer assigned, in recognition of his talent for computational mathematics, to the computing team. There he worked, in particular, on the problem of defining the geometry for the doubly-curved roofs. For the next seven years his time was almost completely devoted to the Opera House.

When a need for a site engineer arose in 1963 he was drafted in to work on the project in Sydney along with a small group of young Arup engineers from the UK. This proved to be a critically influential move for Rice. At Sydney he learned a great deal from the architect, Jørn Utzon, for whom he had the utmost respect and admiration. In the years spent in Sydney he was not simply getting the Opera House built. His hungry mind was taking in all that was going on around him, and his appreciation of what the architect and engineer could achieve together was forming.

He learned from Jørn Utzon about the importance and complex nature of the conceptual aspects of architecture. As the Opera House began to take shape, in which he was part of the process of turning architectural intention into physical reality, he developed a deep admiration for the particular role and skills of the architect. A

Fig. 1.2. Sydney Opera House under construction.

Photo: Ove Arup and Partners/David Messent

Photo: Ove Arup and Partners

Fig. 1.3. The Opera House completed.

respect that he was to remain evangelical about for the rest of his professional career.

After the heady and stressful Opera House project Rice asked for a year out as a visiting scholar. Jack Zunz recalled that he wrote to a number of leading engineering schools in the USA, making it clear what he wanted, and to see which had the best to offer.[6] What he did make clear was that he wanted to learn. Letters after his name were of little consequence. The contents of one letter sent are very telling. One extract reads

'I would like to study the application of pure mathematics to engineering problems. I think that a more thorough understanding of the nature of equations used to solve structural problems in design could lead to a better conditioned solution and ultimately to a better choice of structural components.'

In 1967, at the age of 32, the particular and unique qualities that defined Peter Rice's talents were clearly becoming apparent.

He returned from a year in Cornell University in 1968, and began the process of developing creative relationships with architects who, like Rice, were to receive international acclaim. The first of these was Frei Otto, whom he worked with on what was to become a lifelong interest, in lightweight structures.

13

Chapter One
Rice, engineering and architecture

Photo: Ove Arup and Partners

Fig. 1.4. Les Nuages (the clouds) beneath La Grande Arche, Paris.

Rice's international fame came with the Beaubourg, later known as Centre Pompidou, in Paris. There he worked as engineer to Piano and Rogers, which was the start of a professional relationship with each of them that lasted through the whole of his career. Like the Opera House the Beaubourg building has become an icon, inextricably linked with the parent city. Like the Opera House it has become regarded as a seminal piece of architecture. And like the Opera House it was a major landmark in the evolution of Peter Rice as engineer–designer.

Following the fame and acclamation of Beaubourg he went on to work with other distinguished architects including Norman Foster on Stansted; Ieoh Ming Pei on the glass pyramids at the Louvre; Bernard Tschumi on sculptural structures at La Villette; J. O. Spreckelsen on La Grande Arche in Paris and Adrien Fainsilber on the transparent Bioclimatic Façade, also at La Villette. He worked in different ways on different projects and with different architects. The overriding rule was always this: the role of the engineering is to enable the most exciting piece of architecture to be realised. His collaborators came to learn that if anyone could make their ideas work it would be Peter Rice.

He spent his whole professional career attached to Ove Arup and Partners, moving through the hierarchy from graduate engineer to Director. That brief description suggests a conventional and linear passage through the professional ranks, but that is far from the truth. After the success of Beaubourg he set up an Anglo-Italian partnership with Renzo Piano with the aim of taking on unconventional projects and developing unconventional ways of working. Later he established the still extant practice RFR with architect Ian Ritchie and engineer Martin Francis. As with his link up to Renzo Piano his attachment to Arups continued in parallel with the work of RFR. Again, the goal with RFR was to take on unconventional design projects and through them develop new ways of working as a professional creative team.

Despite his enthusiasm for working in experimental practices he was constant in his praise for Ove Arup, the man whom he referred to as his 'father of engineering'. When Rice joined Arups in 1956 Ove Arup had already got the title of 'The Old Man'. It was Ove Arup's demeanour and attitude to engineering that was one of the reasons that Rice gave for his reason for choosing Arups as the place to try his engineering hand.[7] He admitted frankly that he joined Ove Arup and Partners because he had 'heard it was a place where an oddball could fit in'.

At the time he joined Arups Rice was still not convinced that he wanted to be an engineer. But what does seem clear from the outset is that, in Rice's mind, if he was to become an engineer then his role as engineer was going to be an unconventional one. Ove Arup had created an engineering organisation that was large, flexible and accommodating enough to allow the Rice qualities to develop and the spirit to be fuelled rather than dulled. Rice put it more bluntly. He said that the atmosphere allowed him to 'survive'.

The youthful passion for engineering and the engineer's role in society that stayed with Ove Arup into his late eighties was taken as an example by Peter Rice.[8] In 1970 Arup spoke in the Alfred Blossom lecture of the Royal Society of Arts in terms that were to resonate with the beliefs of Peter Rice.

'...everything built is architecture ...Everything made by man for man's use has to be designed. And in all these spheres dedicated engineers are trying to conjure forth that mystical spiritual quality which is the essence of art.'

Approaching ninety years of age Arup still remained evangelical about the role of the engineer as a concerned and intellectual human agent in society. In a statement to the Fellowship of Engineering in 1983 Arup contended that 'Ideals must be tempered by realism but should not be poisoned by cynicism or hate. In the end all depends on our own integrity'.

Arup's words were taken to heart by Peter Rice and, although he never had the opportunity to have an active working relationship with Arup, those words and the philosophy that produced them stayed with Rice for the rest of his career. They were the foundation on which Rice's design life was constructed. They reinforced the belief that the engineer's approach should be catholic, that engineers should be aware of their duty and potential, both in terms of the wider society and others in the design team.

Reflecting on Rice's career his architectural collaborators are all united in their view that they particularly appreciated his considerable skill at working with and nourishing the architectural idea. They are consistent in the view that with his support the architectural flame would always burn more brightly.

Frank Stella who Rice worked with on the idea for a complex and unconventional roof for the Museum at Groningen described the experience of working with Peter Rice like this

'Once he had a handle, once he could grasp the image, Peter just rolled on like a juggernaut, crushing the obstacles of practicality and cost, making it possible for us to build what we liked.'

And Richard Rogers described him as 'a true virtuoso' who was always optimistic and open to new challenges, always pushing

Photo: RFR Paris

Fig. 1.5. A Frank Stella sculpture.

Chapter One
Rice, engineering and architecture

the boundaries a little further yet *'totally conscious of his professional responsibilities'*.

Contribution to architecture and engineering

Peter Rice as engineer clearly made a significant contribution to architecture. The words of his architectural collaborators, the evidence in his buildings and his RIBA Gold Medal are testament to that.

But from a more philosophical standpoint we should not forget that he played a considerable influential role in enabling a prominent group of architects to realise, refine and evolve a genre of architecture through the late 1970s to the early 1990s, by providing the necessary technical know-how, and the engineering skills that supported the high technology philosophy.[9,10] But he did more than make a contribution to architecture. He made a contribution to all of those in the design team by explaining to them, in a way that could be easily understood, how he thought the structure and the technology could and would work: how the engineering could enhance the architecture.

He also made a significant contribution to engineers. Young engineers who worked with him were, it is consistently reported, working with someone who would take the trouble to explain and interpret complicated or unusual engineering approaches. Rice was someone whose approach was heartfelt, but flexible, shared, but not rammed down your throat.

He was most at home with, and was explicit that towards the end of his career he particularly sought, projects that were envisaged in the broadest sense. Ones which came without the constraining shackles of preconceptions in the minds of the client or his co-workers on the design team. Ones where experimentation and new ideas could be tried. Ones where the constraining conventions could be challenged, to illustrate that there was indeed a better alternative.

Rice regarded himself primarily as a strategist or strategic thinker rather than an engineer according to the conventional Anglo-Saxon interpretation of the word.[11] But being a strategic thinker on its own is not enough to explain the special qualities that Peter Rice developed thoughout his career. One admirable quality was that he could explain his line of reasoning, and his own way of interpreting the problem, to his engineering colleagues. What this meant for young engineers working with Peter Rice is that they not only saw how he worked, but their personal development was also enhanced by this sharing of ways of interpreting problems and understanding how creative solutions could be achieved.

A particular case is quoted by Jane Wernick from the time when she worked as Rice's aid on the BBC Radio HQ project with the architect Norman Foster. There was a meeting at which the front glass wall was being discussed. The wall was to be supported by what was to become a Rice trademark, an array of cables working as a tension field. Such structures are typically complex at first sight and this was something that appealed to another side of Rice's personality: more of this later. Rice told Jane Wernick to imagine that the supporting structure was a bicycle wheel on its side, but using secant ties instead of radial ties. Immediately the way the structure was intended to work was clear and understood, and the calculations could be performed.

The design process is cyclical. Peter Rice thought that it was part of the engineer's job to make sure that it did not become a roundabout. Sometimes there might be a potential for the design process getting stuck if the engineer insisted that something could not be changed; like the size of a structural element being reduced. So he would say that it could, knowing that in all likelihood the design would change in a way that would eliminate the problem, or allow a different solution to the problem anyway.

Peter Rice was eclectic and this eclecticism provided a rich and varied base from which the engineering skills could flow. His inspirations were drawn from a wide range of sources and interests. His approach would vary from project to project. He worked in whatever way he thought would result in the best building. That was the key. The nature of the project, the nature of the relationship with the other design professionals, the desire to try something new and particular personal interests could all impinge on the product and influence the outcome. This was a defining virtue in Peter Rice's work, and the reason why many of his buildings are so innovatory and inspirational.

That virtue, however, makes it difficult to pigeonhole his work so that chapters for a book like this can be easily formed. The work does not fit comfortably into neat divisions by chronology or some other theme. Early ideas would be revisited, refined and reapplied in parts of later schemes. Consequently, the first part of this book deals with the key ideas, themes and influences. The later chapters deal with specific projects or project types, showing how each of the ingredients was mixed in that particular project. But there are many other ways that his extensive catalogue of projects, talks and writings could be arranged. He would have found it both amusing and reassuring that his work is so difficult to sort and arrange. If it had been easier to catalogue his various ventures this would have indicated a predictability and repetition that would have been anathema to Rice.

Photo: André Brown

Fig. 1.6. Model of Bari Stadium: Renzo Piano.

Engineer or architect?

Peter Rice was commonly asked, '*Are you an architect or an engineer?*' or, on some occasions, '*Are you an architectural engineer?*' Indeed, in one of the many complimentary obituaries to Peter Rice one writer suggested that Peter Rice could have been one of the greatest architects of the late twentieth century. But this is missing the point that Rice himself made on several occasions[12,13,14] and one that can be made again here. The world does not need more architects like Peter Rice, what the world needs is more engineers like him.

He rightly identified that the best buildings result from a symbiotic relationship between the architect and the

17

Chapter One

Rice, engineering and architecture

engineer: one where the qualities of the good engineer, those of being an objective inventor, are married together with the more subjective and creative qualities of the architect. It is for this reason that he objected strongly to the term architectural engineer.

Rice was at pains to note that the word 'design' does not exist as such in the French language. The French use of the equivalent word indicates drawing rather than 'design' as it is used in English. Rice preferred the French word *créateur*. Equally, Rice regarded the use of the word

Fig. 1.6b. Rice in France 2: Canopy at the CNIT building, La Défense.

Fig. 1.6a. Rice in France 1: Les Nuages (the clouds) at La Défense, Paris.

Photo: André Brown

Fig. 1.6c. Rice in France 3: Pyramide Inversée at the Louvre.

Photo: André Brown

Fig. 1.6d. Rice in France 4: New end wall for the cathedral at Lille.

'engineer' in English as frustratingly limiting. Having worked in both France and Italy, he was acutely aware that both the title and the implication of the word 'engineer' was significantly less limiting in these two countries than it was in the Anglo-Saxon interpretation of the word.

Rice believed that one of the key issues that engineers should be addressing is that of changing the perception of the engineer's role in the design process. Many engineers regard that role as being to make decisions and calculations based on knowing that there is a substantial body of knowledge,

usually containing several codes of practice, to back up those decisions and calculations. He challenged engineers to break new ground. He was not satisfied with making a structural element as slender as permissible. He wanted to know if a slender structural element contributed to the architectural intention; to a coherent overall solution; to a good piece of architecture.

In his working life Rice acted in three different roles. First, chronologically speaking, was the relatively conventional role as engineer in a multi-disciplinary team. Second was the role as specialist, where he

was given a particular aspect of a project because of his particular skill in working with new materials or in realising conceptual ideas. And lastly was his role as partner in a number of architect–engineer collaborative ventures, perhaps most notably with architects like Renzo Piano. Each of these ways of working produced different kinds of architecture. But in each the engineering is consistent in the sense that the governing principle is to adopt a strategy that is best suited to fostering the architectural intention.

There are many examples that could be taken from almost any of his buildings from the Beaubourg project onwards that would illustrate how his approach would be to work in tandem with the architect and the architectural idea. At this stage one project, chosen almost at random, serves to illustrate the point. In Fig. 1.7 Rice is sketching potential solutions for a new lightweight glass and steel roof to enclose an existing courtyard at the Louvre. Here, he was working with the architect I. M. Pei and the sketches of two of the potential solutions reveal that Rice is working through ideas that take heed of both engineering and architectural matters.

Getting a good quality light into the space is important, so that is figuring in the sketches, one of which shows an oculus (eye) at the crown to admit daylight. Also important is the device to encourage airflow

Chapter One
Rice, engineering and architecture

through the space by some kind of venting system. But the context is a classical set of buildings around the courtyard so a very critical architectural junction occurs where the springing points for the new steel structure meet the top of the existing stone buildings. The sketch shows that Rice intends that this junction should be visible so that the transition from old to new is made clear and explicit, not hidden and therefore ambiguous. There were other subtle touches that add to the quality of the Louvre courtyard structures. Ian Ritchie recalls that

'Peter observed that the maximum member size that we had developed for the Louvre courtyard roof structure had been based upon the diffusion of a shadow line created by an object between the light source and the ground plane as defined by the French mathematician and philosopher Jean Le Rond d'Alimbert (1717-83)'.

Now, I have not read them all, but I would bet that the quality of light and shadows cast by a structure does not appear in any textbook dealing with how to size structural elements. But in a Rice design all of these interrelated matters were given attention synchronously. Rice, as engineer, is not simply settling for a structure that could do the job, he considered alternatives that met both architectural and engineering criteria.

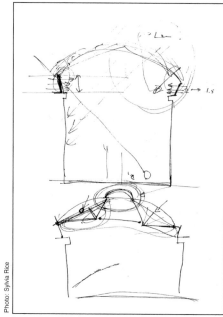

Photo: Sylvia Rice

Fig. 1.7. Rice's sketches for a glass and steel roof over a courtyard at the Louvre.

This philosophy and approach is one that Rice became increasingly passionate about as his career developed. He not only promoted this flexible and enabling role for the engineer but he also positively denounced the engineering stance which got in the way of the architectural solution. He frequently expressed his dismay and exasperation at such attitudes in engineering; attitudes that saw the constant application of apparently rational argument being used to dull, and eventually kill, architectural creativity.

Rice always strove to take the polar opposite approach. He was clear that he did not want to be the architect, but he did want to be the enabling force that gave the architectural inspiration its head. For him the joy was in finding and devising technical solutions that enriched architectural creativity.

He firmly believed that the architect had particular skills that the engineer both could and should not try to usurp. Equally the engineer has particular skills that the architect is not able to replicate. What he argued for was that discussion and interchange of ideas between the architect and engineer should begin at day one of a project and be continuous until the completed project was handed over to the client. And there, really, is the bottom line. Peter Rice was an engineer. As an engineer

Photo: André Brown

he simply wanted two things. First was for the role and potential of the engineer to be properly recognised in the Anglo-Saxon design world. Second was that he wanted to be a good engineer. In the second of these two desires he certainly succeeded. In the first he certainly made inroads that other engineers should capitalise on and extend.

Fig. 1.8. The roof as constructed.

Chapter Two
Themes and influences

Despite the fact that his work is so difficult to categorise and pigeonhole, there are common themes that have driven Rice's work, and these themes became stronger as his skills matured. Those key themes can be summarised as: Materials, Computational Mathematics and Mystery. They were all probably there in the Rice spirit at the start of his engineering career, but the Opera House at Sydney was the place where he could see how inspiration and ideas could be drawn from each of them, and how their potential could be exploited. By the time of the Beaubourg project in 1971 Rice was ready to start applying those themes, using them as foils to the architect's intention.

From Beaubourg onwards he used each of the three in different measure, on different projects. His best projects were always those that gave him the opportunity to explore a new idea in at least one of those themes. The three themes became cornerstones that helped define the aims and facilitate the invention that would be evident in the maturing Rice approach.

Materials

Peter Rice credited Turlogh O'Brien, the materials expert at Arups, with giving him the grounding in understanding the importance of the behaviour of different materials and the inspiration to exploit this as a theme. Calling it a theme is understating the matter. Rice took the idea to the point of evangelism. He observed that certain materials were not generally used in the construction industry and he kept a bank of such materials, their properties and potential applications in mind, ready to be exploited should the right project come along.

When Rice talked of properties he was typically poetic. It was not numerical quantifications that defined materials for him. Instead he talked of their 'character' and 'special nature': the cast steel at the Beaubourg, polycarbonate in the IBM Travelling Exhibition, thin sheets of marble in the French cathedrals and ductile iron in the Menil Collection all fit into this category. All of them are examples where the material is instrumental in defining the architectural quality of the building. Quality achieved in this way is far from being a new idea. Both the gothic stone masons,[15] and the pioneers of new iron and steel structures in the early nineteenth century, as demonstrated in the Samaritaine Building (Fig. 2.1), had, in Rice's mind, a due regard and feeling for the materials that they were using to form the building. They had to understand the material in a way that meant that the skeleton and skin of the building were attended to in equal measure.[16] The results that they left behind were carefully composed objects that were assembled from crafted components.

Rice was particularly concerned at the

Photo: André Brown

Fig. 2.1. The Samaritaine Building, Paris: craft in early iron and steel building.

post-industrial phenomenon which has seen the production of the elements of the building being divorced from the building itself. Consequently, contemporary buildings are complex, that complexity hiding the materials, the technology and, in the end, the architecture. Whenever possible he looked for ways to return to more crafted methods of production, to give buildings a human quality that has been lost in our post-industrial society.[17]

Fig. 2.2. Ductile iron elements for the Menil Museum.

Fig. 2.3. Hand preparation of the ductile iron pieces.

In Rice's mind a particular care for and attention to materials were important devices in the quest to break free from industrial tyranny. He wanted to challenge the prevailing belief that there was little choice in technological issues; that there was some predetermined, underlying logic that we all should follow. The way to challenge the prevailing tyranny, in his view, was to return the evidence of the human back to the designed object. That way, he thought, *'by looking at materials, or at old materials in a new way we change the rules. People become visible again'*.[18]

Broadly speaking there were two ways in which Rice sought to apply and exploit his interest in materials. The first was to respond to the particular qualities and properties of relatively untested materials. The second was to seek new ways of working with established materials. He found potential in both approaches.

Fig. 2.4. Finished ductile iron element.

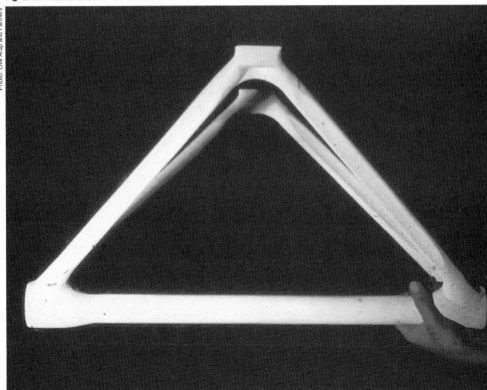

Chapter Two
Themes and influences

Fig. 2.5. Stone, timber, steel and glass composition at Parc André Citroen, Paris.

Photo: André Brown

Responding to a material's particular qualities

One theme that was developed, in conjunction with the collaborating architect, was looking at how the particular properties of materials not commonly employed in structures could be brought into play. How they could be exploited through appropriate detailing, and how the analytical techniques might inform their application and manipulation. He wanted to be able to predict and control the behaviour of the material under its structural load. And the form of the element, its location and its connections would all then be chosen to work in harmony with the material characteristics.

In order to model the behaviour of the material one of his other major interests would play a part. Computational analysis would be brought to bear on these problems.

Much of the work that Rice undertook along these lines was in collaboration with Renzo Piano who shared Peter Rice's passion for working with unconventional materials in unconventional ways. Buildings like Beaubourg, the Menil Gallery and the IBM Travelling Exhibition reveal very explicit consequences as a result of this particular approach to working with novel materials.

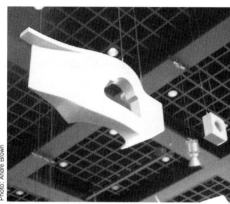

Photo: André Brown

Fig. 2.6 Full scale timber model of parts of the cast steel gerberettes for Beaubourg.

Fig. 2.7 Mounting spring to prevent shock loading in the glass panels at La Villette.

Photo: RFR Paris

26

New ideas for established materials: steel and concrete

Given the established structural materials of steel and concrete the challenge that Rice saw was to look for better and unconventional ways of incorporating the material in a new situation.

Projects such as the Patscentre in Princeton fall into this category.[19] This was seen as a vehicle to demonstrate and make explicit the ability of steel to carry loads in tension. Consequently, the tension became the overriding visual message in this structure and to emphasise this the connections were made light and compact. The tension elements therefore deliberately read as a line rather than part of an articulated assemblage, so the engineering problem was to reduce the bulk of the connection down to the minimum possible.

Conversely, Lloyds of London had to be built in concrete because of fire regulations in the City of London. It is interesting to reflect that despite whatever architectural genius might be brought to bear, it was particular fire regulations in the two capital cities of Paris and London that dictated the height of the Beaubourg and the principal structural material for Lloyds.

In concrete structures, generally speaking, the material is optimised by making joints continuous (not pinned). In Lloyds the challenge was to achieve and

Photo: Ove Arup and Partners

Fig. 2.8. Patscentre, Princeton.

express the quality of the material. That quality had to match the high standard set by the building as a whole and was accomplished through a managed arrangement of precast and in situ concrete in the building as a whole. Here articulation was important so precast and in situ elements were alternated to reinforce that articulation. But the alternation of the two techniques also made it possible to achieve a high quality finish to the concrete surfaces.

So, even here in a concrete framed building, a conventional solution is not accepted. Instead the solution makes the building special, innovative and supportive of the architectural idea that drives the engineering ambition.

Mathematics

We could account for the success of Peter Rice's approach to achieving excellence in engineer–architect collaboration by pointing to his artistic instinct and interests. Though they clearly played a part in helping Rice talk to architects and to interpret their conceptual desires there was more to it than that.

In some ways it would appear that his deftness and confidence with computational mathematics and mathematical modelling might have been hamstringing in discussions with architects. Numeracy and graphicacy in many cases do not go hand-in-hand. But with Rice they did, even though he would only credit himself with a *certain competence* in mathematics. The 'certain competence' that Rice credited himself with is described as supreme competence by others.

John Blanchard was singled out by Peter Rice[20] as the gifted numerical analyst at Arups who showed him how to apply his native mathematical talent in an engineering design situation. It is clear from the way that he talked about colleagues like John Blanchard and Turlogh O'Brien that he valued the fact that he could learn from other engineering colleagues. One of Rice's particular strengths is that he could draw from a range of influences in the arts, culture and science. It is easy to forget that he was equally happy to learn from other engineers.

Tristram Carfrae[21] and Jane Wernick[22] both recount similar stories when they discuss their role as junior (at least in official terms) engineer to Peter Rice. Each of them noted that Rice would say he could calculate something even if he had not tackled a problem of its kind before. Sometimes this might be because he thought the problem would have changed fundamentally before it ever got to the calculation stage. But in other cases it was simply that he was confident in his ability to tackle the mathematics and that meant he could stick his neck out if the potential in the architectural solution warranted it.

His mathematical skill and ability to think in numbers was clearly a talent that was with him from an early age. He said about his youth that

'I knew I loved numbers. They all seemed different, with their own special quality. Each number was precise, it meant something…with numbers I could play all day in my mind.'[23]

A page from his sketchbook dated Jan/Feb 1985, in Fig. 2.9, gives a good indication of the way that Rice would employ a mathematical approach to help solve a problem. Alongside notes and sketches of details and overall structural solutions Rice throws in a matrix calculation to help him understand and explain the tension net that he was trying to devise. For him the drawing of the detail at a critical junction in the net, the sketch of the overall form and the matrix multiplication describing the relationship of forces and displacements are all equally valid and equally important in understanding the engineering in the problem. With an understanding of the interrelationships developed in this way, Rice was able to talk to the architect with confidence and an intimate understanding of the consequences of any modifications.

Rice's dexterity and understanding of the numerical–analytical aspects of a design problem would be with him as a constant support whether he was working in his head, on paper or with a computational analysis. In terms of computational analysis, a key tool in his work was the Dynamic Relaxation program developed for work on early lightweight structures like the City in the Arctic. A particular brand of computational-analytical technique was needed that could cope with the non-linear[24] and large displacement phenomena associated with this kind of structure. Brian Forster[25] notes that this early period of development of the dynamic relaxation technique 'coincided with the presence of gifted engineers such as Peter Rice and Alistair Day'.

In Rice's work he was always seeking to provide architects with solutions where the

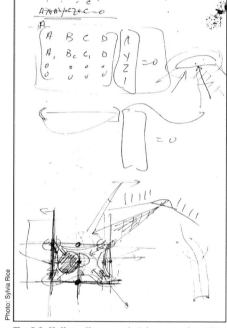

Photo: Sylvia Rice

Fig. 2.9. Mathematics as a sketch: a page from the Rice notebook.

compressive elements could be reduced, where tension and compression could be harmonised, and where delicate filigree forms could be conceived, to produce daring optimal structures. This kind of desire led to a second generation of the Dynamic Relaxation program being produced which now incorporated beam and strut elements. This new programme was used extensively by Rice in his projects. The cantilevered arch ribs at Bari Stadium were made slender by adding fans of bracing rods to increase their buckling capacity. The TGV stations at Lille and Charles de Gaulle Airport optimise the interplay between the structural compositions and the forces applied to them. The arched ribs for the roofs over the Louvre courtyards and Chur transport interchange are made exceptionally thin by being braced in plane by a set of radial cables. All of these are examples of Rice being able to widen the scope, to say yes to an architectural idea; and in some cases even more than that, to suggest more daring leaps of imagination than were being conceived.

Rice likened the architect's job to that of the jockey trying to stay on a difficult racehorse. This is an example of one of his primary interests, producing great buildings, being likened to one of his other prime interests, horse racing. He saw his job as engineer as being able to help the architect to stay on the unpredictable beast. His consummate mathematical ability allowed Rice to hold out the technological hand to support the architectural jockeys and to give them the confidence to stay with what might be a difficult idea.

Mystery (and potential instability)
In buildings like the IBM Travelling Pavilion which Rice undertook with Renzo Piano someone looking at the choice of materials would question the use of polycarbonate as part of the structural make-up of the building. Its structural properties make it a difficult material to use and other engineers would have rejected it on these grounds alone. But for Rice this was part of the challenge and intrigue.[26] How do we use these materials in a way that responds to their particular qualities? The answer would be revealed in aspects such as the detailing. In the IBM building it is in the special joint which gives a structural connection while allowing considerable expansion and contraction. To the person, even the informed engineer, seeing the building for the first time the question would be why is this joint made this way, and how does it function? This kind of reaction appealed immensely to Peter Rice. He would avoid the use of standard solutions and prefer a solution that would make the viewer look and think.

The intrigue that Peter Rice sought would often arise from his interest in materials, but it could equally arise from the choice of structural system. At La Villette the tension structures supporting the glass walls of the Grandes Serres are difficult to understand on a quick inspection. Unlike the standard beam or truss where the function and structural actions are relatively clear and explicit, in these tension trusses the structural actions take some working out. This is partly because the way the structure works changes depending on the direction of the wind pressure. It is also partly because the compression elements are taken down to a minimum, in both number and size.

But one of the very particular ways in which Rice created mystery and uncertainty was in playing with the idea of stability. He was best able to engage with this idea in tension assisted structures, and this, to some extent, explains his fondness for such structures.

When he talked about the tension structure supporting the roof of the Usine multi-store at Nantes, in western France, he was unrepentantly clear that he aimed to produce a building which felt 'ever-so slightly unstable'. The mirroring of actual tension by using visual tension was appealing to Rice, and, anyway, he found the idea that the building looked like it might fall down entertaining.

The feeling of apparent instability could

Fig. 2.10. Apparent instability: a mullion truss at Parc André Citroen.

Photo: André Brown

Fig. 2.11. Challenging convention: a reversed bowstring truss at Charles de Gaulle TGVR.

Photo: André Brown

be created in one of two ways. It could be that the element or building looked unstable because stability was provided in an unusual way. Or it could be actual instability: an instability that was only corrected when the building or element moved and certain elements, usually tension cables, suddenly snapped into action.

...and Mastery

The subtitles of this section are Materials, Mathematics and Mystery. These words capture in abbreviated form much of what Peter Rice was about. But to complete the alliteration we could add a fourth M: Mastery. Because having an interest on its own is not enough. Peter Rice had both the ability and the confidence in his ability to turn interests into tools. Tools that he could use to work on a building and in so doing transform it into an object that was particular and special.

In all of his significant buildings it is these qualities that define what is special. They augmented the architect's skills and provided his co-workers in the design team with a rich palette of alternatives and potential. Perhaps, most importantly, he provided the architects with an ally who gave them the confidence that the difficult could be achieved and that there would always be an engineering solution to a good architectural idea.

All of those who worked with him on his most significant buildings testify to the fact that it was this aura, backed up by a consummate engineering ability, that allowed them to develop their ideas free from the shackles of tried, tested and, certainly in Peter Rice's mind, staid solutions.

Inventiveness and reflection

Later in Peter Rice's career he was so busy dealing with the wide array of projects and problems that were pushed in his direction that he found less and less time to devote to reflection about his work and what he was aiming to achieve. Or, at least, that is what he said. In fact, he had numerous invitations to talk and write about his work. For those cases where he was able to devote the time to prepare a talk or paper what is significant is that he was always incisive and thoughtful. He did this three to four times a year; mainly to audiences with a strong engineering bias, where, he thought, the seeds of his gentle evangelism might prove to be most productive.

Despite what he might have said publicly, he did enjoy the fame and praise that his work post-Pompidou attracted. But it is typical of the man that he did not use occasions of public writing and public speaking to simply sing his own praises and laud his own projects. He saw such occasions as places to share his motives, to crystallise his thoughts and to invite stimulating comment. And if the responses that came were born out of thoughtfulness, consideration and a desire to be inventive, challenging and innovative it did not matter whether that comment came from the most junior of students or the most senior of his peers, they were welcomed equally.

For Peter Rice it was not the status of the question bearer that was important, it was the quality of the question. It is clear from his observations, and the comment of those who worked with him, that what he found particularly frustrating and stifling were views that were derived from the standpoint of someone who wished to preserve the cosy comfort of the status quo. He found attitudes that can be characterised as '*this is what we did last time so let's do it again*' or '*this is what the code of practice suggests*' not simply unappealing but downright offensive.

Rice saw a clear division of roles between the architect and the engineer, and was vitriolic about engineers who claimed to have a foot equally in both camps. In this respect a particular focus for his uncharacteristic venom was Santiago Calatrava (see later). In seeing the division Rice was not handing over the creative part of the game to the architect. The opposite is close to the truth. Rice was passionate about the need and potential for the engineer to be inventive. His ideal world promoted the symbiotic relationship between the architect and engineer and saw the interplay between the two as the very basis for good architecture and good design.

At a seminar held in the south of France called 'Art and Technology, the East and West' a variety of disciplines including artists, architects, journalists and Peter Rice, as the engineer, met. At that meeting Richard Weinstein, who was head of the architecture school at UCLA, commented that the problem with engineers is that they are '*Iagos*'.[27] Put simply his argument was this. Iago is the character in Shakespeare's *Othello* who kills romantic and sensitive aspirations by reiterating rational argument. Iago, as the agent of rationality above all else, succeeds in breaking the fragile ideas that are based on passion and flights of fancy. In the end heartfelt passion is denied by building a barrier of pragmatism beyond which it is impossible to negotiate. In the same way the engineer can kill the essence of creative design ideas by arguing that rational and pragmatic considerations make the architectural proposals unachievable. Iago is thus the prototype for scientific man, as described by W. H. Auden in *Joker in the Pack*. 'Scientific' is used here in the worst sense, to indicate the eradication of art and creativity by the application and reapplication of ruthlessly inflexible logic.

This analogy, the (bad) engineer as

Photo: André Brown

Fig. 2.12. Model of roof structure at Kansai Airport: with Renzo Piano.

pragmatic solution or a wild conceptual image. There was no place for Iagos in Rice's vision of the world of architect–engineer collaboration.

Iago, appealed to Peter Rice and he used it on a number of occasions to express his concerns and account for his motives in trying to encourage the engineer to work as an active partner in the creative design team. The source of the analogy being an architect is notable too. Rice is well known for his collaboration with architects like Richard Rogers, Renzo Piano and Ian Ritchie but wherever there was an architect or other member of the design team with a reflective observation to make Rice was keen to hear what they had to say. If what they had to say had the potential to enrich the design then the comment was welcomed whether it was to do with a

Chapter Three
Sydney Opera House

Chapter Three
Sydney Opera House

Photo: André Brown

Fig. 3.1. The Opera House today.

The project, the lessons

The Sydney Opera House was, from the day that the jurors chose the project as the competition winner, to the day the last site worker left the site, steeped in political and professional wranglings. There are several accounts, notably that by David Messent,[28] of the story of the painful birth which led, among many other things, to the father, architect Jørn Utzon, having to desert the child before it was born. But much of the story is an unnecessary diversion in accounting for the development and application of Peter Rice's talents, so this account deals principally with the issues that relate directly to Peter Rice.

One of the most important aspects as far as Peter Rice was concerned was the influence of the original architect, Jørn Utzon, on his way of thinking about architecture and technology. Like Ove Arup himself, Rice was consistent in his view that Utzon was a gifted talent who deserved the utmost respect. Rice worked as site engineer in Sydney for three years and his close contact with Utzon was of great importance in establishing Rice's career-long view that the architect provided a vision and set of skills which the engineer could not, and should not, try to replicate. Instead they were skills that should be supported and responded to.

There are so many aspects to the Sydney Opera House project that it is impossible to say which team was responsible for the engineering solutions. Major parts of the scheme were devised, calculated, refined and drawn only for the basic concept to be scrapped. The costs, the problems of trying to realise the design intention, and the politics were all national news stories. But what finally came out of the labour ward was a child that has quickly matured to become, possibly above all other competitors, an icon to represent the Australian nation. It is one of the few instantly recognisable buildings from around the world.

So, what was Peter Rice's role in the Opera House, and what was the role of the Opera House in forming the ideas that were to be cultivated in the Peter Rice mind? One of the most important answers is that this was the project that established in Peter Rice's mind the notion that he wanted to make a career out of the business of engineering.[29]

Rice was working as a young engineer in the Arup archipelago in London as the Opera House project was being developed. He was working on the computational aspects of the project, developing an adeptness with computers and software that was to be utilised and explored throughout the whole of his later career. The particular problem that he worked on was in trying to devise a way of describing, using computational mathematics, the three-dimensional curved form of the gull's beak shell roofs. These were to be constructed using doubly-curved precast concrete elements.

The breakthrough came with the realisation that the 'shells' could be formed from sections of a sphere. Since the sphere has a very simple regular geometry, describing the shape of each element that defined the outside surface of the shells (tile lid) and the maximising of repetition in the precasting plant became considerably easier. Thus it became possible to produce the

Fig. 3.2. Drawing showing how the roof was subdivided.

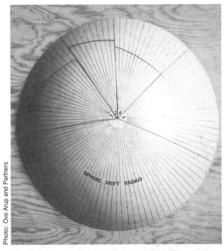

Fig. 3.3. The final geometric solution: taking the forms from segments of a sphere.

Fig. 3.4. The shell ribs starting to take shape.

Fig. 3.5. Precast concrete section being loaded for transport to site.

Chapter Three
Sydney Opera House

Photo: Ove Arup and Partners

Fig. 3.6. The tile lids are fixed to form the outer skin.

high quality, well-detailed elements that would be essential to minimising costs, ensuring weathertightness and achieve the overall appearance that was required.

As mentioned above, it is difficult to account for whose ideas were being applied in a project like this that involves so many aspects, so many individuals and so many teams. But Rice says that, although he was a junior engineer working in the computational analysis team at Arups, he 'did the geometry'[30] of the roof so it appears that he did have an important role in this crucial aspect of the design of the Opera House.

It is worth noting in this context that Rice was not one for claiming ideas when they were not his. In the rounds of praise that followed the successful completion of

the Beaubourg project Rice's key contribution was recognised by his collaborators Renzo Piano and Richard Rogers. Rogers suggests that one of Rice's principal contributions was in having the idea of the gerberettes. But Rice said that the idea of introducing the gerberettes was not his, but that of their young German engineer collaborator Lennart Grut. Such honesty leads to the conclusion that Rice did play some part in resolving the major problem of the geometry of the Opera House shells.

He typically understated his skills in computational mathematics, describing his reason for being involved with Sydney as being because he had a 'certain facility with mathematics'. The truth is that he was a gifted mathematician who had taken easily to the relatively new computational techniques that were introduced at that time. He was at ease with matrix methods which, although used routinely now, were then a novel technique in structural mechanics; a technique that relatively few engineers could handle with Rice's dexterity, if at all.

Whatever the nature of the Peter Rice input to the problem of establishing a neat solution to the Opera House the way that the solution worked became a recurring principle and strength in Rice's work. Peter Rice always said that, despite a rather barren background in terms of inspiring buildings from his youth, one thing that was clear

Photo: Ove Arup and Partners

Fig. 3.7. The tile lids are fixed to form the outer skin.

from an early age was his love of numbers. This love was not for mechanistic calculation but more in applying theoretical techniques to practical situations. That coupled with a particular skill for seeing the patterns and implications behind the numbers was where he saw the delight and the reward.

Here, in the Sydney Opera House project, was an apparently complex problem in which the key was to spot the potential for an elegant solution; to identify some kind of order in the face of apparently unstructured complexity. It was the enthusiasm for seeing a pattern and order in a mathematical description of a problem that not only appealed to Rice, it became a source of innovation for his work. This particular feature of Rice's character was reflected in Rice's later interest in chaos theory.

When we look back at the evolution of the design of the Opera House shells it is now apparent that some of the qualities that Peter Rice brought to the project were to become evident as a hallmark of his particular approach and problem solving capabilities. The Sydney Opera House building was a project that came to Peter Rice early in his career through a combination of particular circumstances. But Jack Zunz's decision to send him to Sydney brought out the best in Rice. On site

Photo: André Brown

Fig. 3.8. Springing point for the arch ribs: a crucial element in the site operations.

Photo: André Brown

Fig. 3.9. The junction at the crown takes on a powerful visual role.

he was the resident engineer, and the practical problems of getting the construction completed were dealt with as effectively and efficiently as he dealt with problems of mathematical analysis.

The roof structure at the Sydney Opera House encloses the two main concert halls and a restaurant. The dominant elements that are most closely associated with the Sydney Opera House are the gulls beak shells whose strong three-dimensional geometric form have become synonymous with the city of Sydney. But, in fact, the roof is made up of three major elements. In addition to the main shells, there are side shells and 'louvre' shells. These final elements derive their name from the louvre walls that they replaced.

The height of the largest shell from its springing to the crown is 54·6 metres. It is 57 m wide and the smallest shell is 22 m wide. Like all of the shells the large shell is symmetrical about a central axis which coincides with the centre line of the hall below. The hall length, measured from the ends of the shells above, is 121 m. Each shell is, in fact, a mirrored pair of ribbed half shells which meet at a ridge. The ribs that run through the shells were, in general, made in segments and the longest of these ribs was fabricated from 12 standard segments from springing to crown.

The form of the shell had evolved over

Fig. 3.10. Careful attention to detail achieves weathertightness and visual integrity.

Fig. 3.11. The Opera House and Sydney Harbour Bridge.

Photo: André Brown

a period of years. When Rice arrived at site to work with the contractor, Corbett Gore, he was well aware of how the final solution had been conceived and how the final geometry, based on taking sections from a single sphere, should work. That knowledge and his extraordinary competence with computational mathematics made him a key player in turning the dream that had been in danger of becoming a nightmare, into reality.

John Nutt recalled the time when it came to setting out the main ribs for the shells. These ribs were made in segments at a dedicated compound close to the site and transported by lorry to the site (Fig. 3.5). It was crucial that the springings and the first few segments of the arch ribs were oriented correctly in 3D space. If there was a minor error here, by the time the ribs met the main curved support beam that ran down the centre line of the structure at the crown, the error would be greatly magnified. The correct orientation was crucial.

The surveyors at the site were using conventional equipment and conventional setting out methods. They were baffled by the rib and tile geometry and did not know how to set out the location points for the rib sections. Rice went away, wrote a short computer program, ran the data through and came back the next day with a sheet of

setting out points that the surveyors could work with. To others computers were boxes of electronic stuff that they had a vague inkling about. To Rice they were a useful device that could help him sort out an order in the jumble of numbers that would be floating around in his head.

The less obvious influences

In most reports it is the problems, the evolution and the final solution to the concrete shell roof structure for which the Opera House is most noted. But if we leave it at that there are other aspects that would be missed if we are to understand where Rice found sources of inspiration in Utzon's masterpiece.

The glazed end walls of the Opera House are supported by a structure that spans considerable distances close to the centre line of the shells. The structure on the inside of the plane of glazing is in the form of a compound beam; equivalent in action to an I-beam. The flanges of the I-beam are made of small tube sections and these are welded to a thin, solid steel sheet that forms the web. They span approximately vertically, but actually rake at a range of angles to form a faceted curved sheet of glazing as a whole. The depth of the supporting beams varies in response to the bending stresses; getting deeper towards the middle of the span.

Chapter Three
Sydney Opera House

Photo: André Brown

Fig. 3.12. Glazed end wall.

Photo: André Brown

Fig. 3.13. Raking ribs of steel composite sections support the glazing at the open ends.

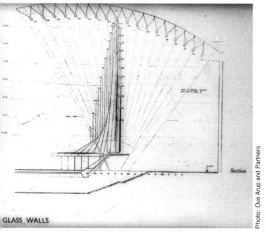

GLASS WALLS

Fig. 3.14. Jørn Utzon's drawing of the glazed end walls.

Photo: Ove Arup and Partners

Fig. 3.15. Glazing detail.

Figure 3.13 shows the raking glazing supports and Fig. 3.14 shows Utzon's drawing of the end walls. Extending from the flange of the supporting beams is a special connector that picks up the glazing. Fig. 3.15 shows the connector and its junction with a restraining bar in which the joint had to be designed to accommodate the restraining bar at a range of angles, as the angle of the glazing changed.

It is interesting to look forward to projects like La Villette (chapter six) and the Railway Station at Paris, Charles de Gaulle (chapter ten; Fig.10.15) and to realise that, perhaps surprisingly, it is the Sydney Opera House where the seeds for these projects were sown in Rice's mind. The way that steel and glass are married together; the invention of new details to give a glass–steel connection and angular flexibility; the way that the supporting structure rakes at different angles to be read as rippling waves from certain directions: these are characteristics which give the ancestry away.

But even in the concrete elements there is much more than the shell roof structures to appreciate and inform. The beams over the entry staircase and concourse are reinforced concrete. But Utzon gave them care and attention he did other parts of the project. Figure 3.16 shows the ends of the beams as they project into the main space above the entry stairway.

Fig. 3.16. The end section of Utzon's 'Moment Beam'.

Photo: André Brown

Like the compound beams that support the glazed walls these beams change along their length, responding to the different jobs that they have to do in different locations. They are deepest in the locations where the bending moments are greatest, hence their name, and they taper off towards the cantilevered ends where the depth becomes minimal. The transitions between the cross-sectional shapes are made smoothly, again reflecting the shape of the bending moment diagram to which the beam profile responds. These structures are, in fact, like a lesson on bending stresses and deflection drawn in the air.

What all of this shows is that the Opera House was much more than a big project that Rice worked on. He was there for a reason. That reason was a particular set of skills but once there he made the most of what was on offer. Close inspection of the Opera House reveals why Rice was so respectful and grateful to Utzon. Rice referred to Arup as his engineering father. It would not be stretching a point to say that Utzon was his architectural father.

Between the Opera House and Beaubourg

Following the completion of the Opera House, according to John Nutt, the team of young engineers who had worked on the project *'were not much good for anything else'*. This is an obvious exaggeration. But it is certainly true that the problems in Sydney were very particular. Those involved did need to see what it was like back in the real world and collect their thoughts after their extended period in the rarefied atmosphere of the Opera House project.

Jack Zunz was much praised for his management of the difficult position of the young engineering team in Sydney. He had deflected much of the political pressure away from the site team, and this had allowed the team of enthusiastic young engineers that Rice was a part of to get on with the job in hand. Zunz extended this avuncular role beyond the end of the Sydney project. The building might have been finished but the development of these young and valuable engineers still needed attention. Each of the team was directed in an appropriate way and for Peter Rice this meant going to the USA to do a Masters at Cornell. This gave Rice time to think and reflect about his experiences at Sydney; his work with Utzon; the potential for the exploitation of materials in unconventional ways; and to assess how his particular skills could be employed in this industry that he had decided he would dedicate his life to.

After returning from Cornell, over the next few years at Arups, there was an involvement with more conventional projects but Rice picked out the work with Frei Otto as being particularly significant in terms of

Chapter Three
Sydney Opera House

the approach that he developed later. Rice identified the Frei Otto project for the Arctic City as influential in the development of ideas relating to scale and materials.[31] But just as important about this competition project was the fact that the way that the professional relationships worked and developed was in an unconventional manner.

The Arctic City was to be housed under a dome of 1 km span; so, it is immediately obvious why scale is an issue. Likewise the materials had to be chosen with care. Rather than restraining the large inflated dome with steel cable other materials such as polyesters proved to be more appropriate. Such materials are more flexible than steel allowing for the considerable thermal expansion and contraction that was going to be experienced by the restraining elements of the dome. This, in turn, generated particular kinds of solutions both at the level of overall structure and at the level of detailing.

Again, another seed was planted. Rice was intrigued by the relationship between the element being designed and the material from which it was appropriate to make the designed element. This link was to become a core theme in many later projects.

Through the Frei Otto projects Rice was able to appreciate the real potential in new ways of collaborating in interdisciplinary teams. This was to fuel his desire to such an extent that in the Lloyds building which came later he was to feel somewhat hamstrung by conventional professional roles. Looking back, these can be seen to be the days when the seeds of ideas that would lead to the establishment of his experimental practices Atelier Piano and Rice, and later RFR were being sown.

Perhaps the industry would be improved if all young engineers could be given this kind of project on which to work. Ones where the roles are looser and there is the potential to gain an understanding of what the collaborators in the process are trying to achieve, and how they are thinking. John Nutt, one of the young engineering team who worked with Peter Rice on the Opera House, certainly believes that the architect 'brings something different, something special to the table'.[32] In part John Nutt thinks that in his case this has its roots in working with architects as an undergraduate. But it is salutary to note that he was the only one in his year who chose to work with architectural undergraduates when given the opportunity.

Chapter Four
Beaubourg

Chapter Four
Beaubourg

The engineer as designer

Piano and Rogers were a pair of young and relatively inexperienced architects when they won the Beaubourg competition against 680 other contestants. Although Ted Happold was the senior Arup engineer for the competition entry, Peter Rice became the principal Arup engineer for the project as the design was modified from the competition entry to become the scheme that was actually built.

Rice had worked on the Sydney project for seven years, and it was three to four years after his participation with the Opera House that he became involved with the Centre Pompidou project. In the meantime

Fig. 4.1. Beaubourg/Centre Pompidou shortly after completion.

Photo: Ove Arup and Partners

Photo: André Brown

Fig. 4.2. Beaubourg/Centre Pompidou in 1999.

he had been at Cornell University as a visiting scholar where he had time to reflect on the lessons that Sydney had to offer. He even reconstructed parts of the Sydney scheme as part of his investigations at Cornell.

In the Beaubourg scheme he was able to begin to crystallise his own thoughts as to what was important as far as playing a creative role in the design process went.

Back at Arups in the late 1960s and early 70s Rice worked in design teams on around ten major projects but, either because of his position in the team, or the nature of the project, he was not able to apply the ideas that he had fermenting in his active mind. Beaubourg was exactly the right kind of project to come his way at that time. It was a project where he was working with young architects who were not tied into a fixed set of professional relationships, on an iconic project where it was appropriate for art and technology to be seen to be working in harmony. This was the kind of project for which Rice had been waiting. It was made for him.

Beaubourg was the project that Rice described as being the point at which his *true design self* was realised: when he finally appreciated the potential of being regarded as a designer, and where he had the means to exploit his interest in materials. Despite the fact that this project came many years

after Sydney Opera House, Rice was candid that lessons learned in Sydney were reinterpreted and reapplied in the Beaubourg project.

On the face of it this is difficult to appreciate. Sydney Opera House, the precast curved concrete shell with a 1950s ancestry; Beaubourg, a 1970s high technology number in steel and glass. Now, one reason why Rice quotes the Opera House as inspirational and a source of ideas for his later projects could be that Sydney was the only project of major renown that he had worked on prior to Beaubourg. After Sydney he had served his time working as an assistant engineer on projects like the Crucible Theatre Sheffield, which, although substantial, were not buildings that received international acclaim. But there is much more to the reference to Sydney and Utzon than that.

So, what elements of the engineering and architectural design languages did he add to his vocabulary in Sydney? The first would be the importance of detail and scale. A care and interest for the interaction of the building and its component parts with the human scale, the concern for the look and the feel close up, were inherited from Sydney.[33] It was scale that Rice quoted as being a key in devising an appropriate engineering solution. How, he wondered, could the human scale be dealt with so that the building would be read as tactile and unintimidating: something with

a human quality? This was to be a building for the people after all.

Although there are engineers around now who do appreciate and understand what Rice meant when he talked about matters such as humanity, tactility and scale, they are still in the minority. Perhaps the nature of the industry makes this inevitable, but Peter Rice would not agree with that contention. For him the engineer's role was as much about optimising the quality of the building as it is the architect's role, no matter what the building was for or how restrictive the budget might appear. Like Ove Arup he firmly believed that engineers should always be aware of the impact of their work on the individual and society. They should think about the consequences of their actions and decisions.

From the outset Beaubourg was to be a steel structure. One of the problems with this is that if we say that a building is steel framed, to many in the industry, certain preconceptions immediately spring to mind about how such a building might look and be made. One of the weapons now established in the Rice armoury was the idea of challenging such preconceptions. At Sydney, Jørn Utzon had challenged conventions such as that of the simple beam, and had developed the idea of the 'moment beam' which responded better in its form to the job it had to do than the conventional

Chapter Four
Beaubourg

rectangular beam. Rice had listened to Utzon and had learned.

In the Beaubourg project, Piano and Rogers were challenging the conventional idea of what a museum was so, as a consequence, Rice was presented with the opportunity to challenge the preconceptions about what a steel framed building was, and how it might look. In forming an argument to support this approach he often referred to the historic precedents that surround us, that can give us confidence and inspiration to challenge convention. Many cities, and Paris in particular, have iron and steel framed buildings which are successful not only in overall terms, but also at closer examination at the level of detail. The Samaritaine building, on the banks of the Seine, is a good example of this genre of buildings to which Rice paid due respect. Such buildings have a quality that Rice referred to as 'pleasing to the eye'. The international exhibitions held in Paris in the nineteenth century had in them many examples of iron and steel framed structures that achieved this quality of being both pleasing to the eye and structurally innovative.

The Victorian engineers, Rice noted, had managed to succeed in ways that are rarely accomplished in contemporary engineering. They were presented with new materials like cast and wrought iron, in a context where the rules about how they were to be designed had not been formulated. As a consequence they had to invent and innovate. They had to work from first principles and trust their own judgement; a judgement that had to be formed from an intimate understanding of the nature and qualities of the material with which they were working. Rice admired the products of this period in which new materials like steel and glass were introduced, but where the crafted and personalised quality of the product remained.

Buildings like the Grand Palais in Paris drew Rice's praise. It was a building that he regarded as an exemplar; a building that illustrated clearly what we have managed to lose. This led him to comment

'We have learned so much about steel and glass and how structures work since then. Where has it gone? Has it become smothered by industry and a desire to standardise? I believe so.'[34]

The close-up, tactile quality of materials was something that Rice was keen to embrace in the Beaubourg project, and he saw in Beaubourg the potential to re-establish an element of craft in the building industry as a means of achieving the quality that he so valued. In this regard he was clearly working with like minds. Piano was an avid sailing enthusiast who had built his own boat and was enthralled by the optimised technological kit that is associated with sailing. He regularly makes carved wooden models that are pleasing to touch and feel. His feeling for the importance of crafted products and experimentation with materials was as acute as that of Rice. Reflecting on Beaubourg and similar projects, Piano observed that 'I have matured by remaining faithful to a permanent underlying thread in my work: the art of making'.[35] The quote would have been equally true if it had come from Peter Rice.

Rice had called at the 1970 World Fair site in Osaka while on a visit to Japan shortly after hearing that the Beaubourg competition had been won. There, in the buildings still remaining, he had seen steel castings being employed and this material was to become a key part of the Beaubourg repertoire. The kind of approach adopted here was also to become a principle that Rice would adopt on future schemes: looking to employ new materials or to find new uses for established materials. This was a key part of Rice's strategy for challenging convention or adding the craft and novelty that could contribute to making a unique building rather than just another building.

Craft and innovation in the individual elements of the Beaubourg structure were therefore going to be two of the ingredients in defining the quality. Breaking the scale down by subdividing the component elements was to be another.

But the structure entered for the competition scheme did not meet these intentions. As originally proposed by the Ted Happold team at Arups, the steel frame was conceived as a one way spanning truss, with long main span and a short span at the outside edge. The idea was that the central core needed to be an extensive column-free floor space. The proposed main span was 48 metres. Outside this was a 6 m wide zone for circulation. The competition winning structure had a line of columns on both the

Photo: Ove Arup and Partners

Fig. 4.3. The structural system proposed in the competition winning scheme.

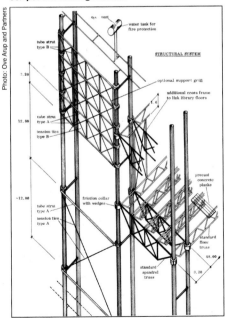

inside and outside faces of the circulation zone that defined a kind of tall colonnade on each face.

It is obvious from the competition drawings that to Rogers and Piano what was clearly important was that the front façade should be read as a transparent wall through which all of the main public circulation should both take place and be seen to take place. This was a cultural factory and should be seen to be so. Changing the structural system that had been proposed in the competition scheme was essential to achieve this. The outside face columns in Arups technical submission (Fig. 4.3) would read more strongly than they had been depicted and the idea of the transparent façade would be lost. The columns were in danger of becoming the bars on a prison window. They had to go.

The solution to the problem was found by replacing the short outer trussed beams that were originally conceived as simply spanning between the inner and outer row of columns. The idea was that in their place would be substituted short gerber beams, referred to as gerberettes. These acted like seesaws with a hinge at the inner row of columns (Fig. 4.5). The main beams would push down on one end (the short end) of the seesaw causing the longer, outside leg of the seesaw to lift. The consequence was that instead of the outside row of columns having

to hold the trussed beam up, they now had to hold the end of the gerber beam down, against the potential upward rotation. Since they were now ties, rather than columns they could be made from thin steel rod or cable, rather than large hollow section. The visual intrusion was thus minimised and the transparent face to the outside world was accomplished.

The Piano and Rogers scheme concentrated the building mass on the Rue de Renard side of the site creating a public plaza in front of the building to the west which is sloped to face the front façade of the Centre. The building is six storeys high with 3 m deep, 48 m long trusses spanning from front to back between the inside rows of columns. The trusses now spanned on to the inside face of the gerberettes along a line that is inside the line of the columns. The gerberettes themselves are pinned through the columns and the tension rod that ties the end of the gerberette down runs from the top floor gerberette, through all of the other gerberettes which are pinned to the same column, and down to the ground where they are tied to a foundation anchor.[36,37]

Although Rice was sometimes credited with the idea of introducing the gerberette solution he confessed candidly that the idea of adopting it was not his but arose in a design meeting and was probably rightly attributed to the German engineer, Lennart

Chapter Four
Beaubourg

Grut. Even the suggestion for the general shape of the gerberettes was not Rice's. He gave the credit for that to the young engineer on the project, Johnny Stanton, whom he described as a talented designer. It is typical of Peter Rice that he gives credit to a junior engineer, and a collaborator in this way. If the idea was good it did not matter where it came from, it was valued and its source was recognised. Rice's particular contribution was the decision to make the gerberettes from cast steel; an unusual material in building structures, but a material used in other industries.

When Rice talked about the gerberettes he said that he preferred to call them '*pieces*'

Fig. 4.4. Model showing the revised structural system.

Fig. 4.5. The gerberettes in the final structure.

rather than elements or parts of the structure.[38,39] This was because he said it made him feel like an artist. They meant that the building was '*liberated from the standard industrial language*'. The word suggested craft and care, not industrialisation and repetition.

So, the idea to use small gerber beams to lighten the outside face of Pompidou was

Fig. 4.6. Model of the gerberette: Renzo Piano.

Fig. 4.7. The gerberettes shortly after casting.

not Rice's. But to make them from cast steel was. This immediately raised an engineering problem: that cast materials are rarely used for the principal load-bearing elements of contemporary structures. There are no obvious codes of practice to use as an easy crutch. As a society and as a building profession the soft answer would be to avoid such circumstances. For Rice, though, this was a situation that was liberating rather than constraining. It provided an opportunity to give *personal design philosophy full rein*.

Cast materials are, by their nature, flawed. These flaws lead to failure. Hence, the challenge in Beaubourg was to find a technique to verify the design of the gerberettes in a way that would satisfy the French building authorities.

The complex mathematical analysis, that was needed to prove these elements, was derived from techniques that were used in industries outside structural engineering; the techniques used in fracture mechanics in the design of pressure vessels and aircraft frames. The appeal of the gerberettes to Piano might have been the crafted nature of their manufacture and appearance. And Rogers might have been drawn to the combination of lightness on the external façade coupled with the clear articulation of the elements of the structure. But for Rice the particular innovation was to apply the new analytical techniques that meant dealing with computational mathematics. Cast steel was not completely new, but using this kind of analytical technique was.

The consequence of using the gerberettes is not only that the outer face was lightened and the transparent outside layer thereby achieved. To Rice what was

equally important was that the bearing, where the gerberette sits on the column, is relatively small. Because that joint occurs at what is such a visually powerful location it plays a major role in defining the scale at which the building is read. The result is that, despite the fact that the building is massive in overall size, it is legible at the human scale. He was much pleased that an old French woman visiting Beaubourg in the period just after its construction remarked that she liked the texture of the gerberettes.

Being cast, the gerberettes have a crafted quality that Rice sought to bring to buildings wherever possible. They are cast and then fettled by hand so that each has a

textured and slightly undulating surface, in contrast to the flat machine finish of the common rolled section. In addition, the cross-section changes throughout the length of the gerberette in a way which is simply not possible in a rolled section. Like the moment beam employed by Utzon at Sydney, the gerberette uses material in a way that responds to the loads being carried. Near the tension rod end (section EE: Fig. 4.8) of the element the bending moments are relatively low so the section is tapered down to a minimal depth. Nearer the support, at section DD, the bending moment is higher so the I section becomes deeper. In contrast, at the column pin and the truss

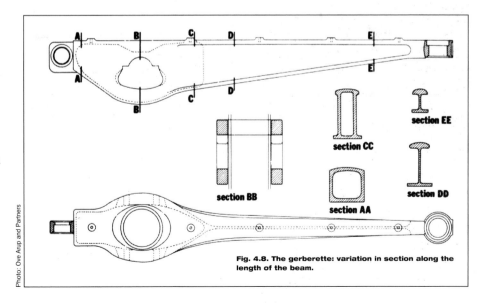

Photo: Ove Arup and Partners

Fig. 4.8. The gerberette: variation in section along the length of the beam.

Chapter Four
Beaubourg

connection point (sections BB and AA, respectively) the shear force is very high so more material is needed in the web. Consequently, the cross-section changes here to a box shape, with two webs rather than the single one of the I section. This massaging of the cross-section into appropriate forms is one of the potentials and consequences of using a cast material. The alternative would have been to weld a range of sections together and, apart from this being a less elegant solution, it would have meant that the gerberette being read as a single sculptural piece would have been much less likely.

And in all of this it is easy to forget that the gerberettes were not the only unconventional piece of technology that can be pointed to at Beaubourg. The columns were also cast, unlike most circular hollow sections that are fabricated from steel sheets, which are originally flat, then bent and welded. For these columns the molten steel is formed into a cylindrical shape by a centrifugal force; centrifugal casting is hardly run of the mill either.

Rice[40] described Beaubourg as analogous to a book that can be read at different levels. At the first level the columns, beams and joints are seen as the elements that constitute a coherent overall structural system. Next the distinction between tension (in the slender elements)

can be clearly distinguished from the compressive system (large diameter hollow sections). Finally, the connections read as families of elements; the egg-shaped holes that receive the pins supporting the gerbers, and the swellings at the end of the gerbers to pick up the main tension rods are two examples of these families. The gerberettes rest on pins that pass through the column, but the gerberettes themselves do not touch the columns. There is a deliberate space between the column and gerberette, like the blank page at the end of a chapter or a line return at the end of a paragraph.

But it was not all plain sailing for Rice during Beaubourg. As at Sydney the politics of the situation were rather fragile, as were the first gerberettes to be cast. Both of these aspects are worthy of note since they reveal important aspects of the character of Peter Rice. There are other aspects, too, that reveal qualities confirmed in his later work.

The first thing that is clear is that he did not take all the credit for a project that often has only his name as engineer attached to it. He picked out individuals that he worked with like Lennart Grut, Johnny Stanton and Andrew Dekany, an engineer of the old school '*an engineer as engineers ought to be*'[41] for individual praise. But he was equally quick to recognise the importance of the team at Arups who provided the engineering support. He openly applauded

Photo: Ove Arup and Partners

Fig. 4.9. Articulation and layering of the constituent elements.

Fig. 4.9a. Articulation and layering of the constituent elements.

Photo: Ove Arup and Partners

Photo: André Brown

Fig. 4.10. Exposed corner: the ends of the trusses integrated into the bracing system.

the actions of colleagues like Ted Happold and Povl Ahm who, like Jack Zunz at Sydney, worked at a high level in the Arup hierarchy to 'create the space for the team to get to work'. Equally he praised his client, Robert Bordaz, who, in the face of criticism and doubt from politicians and writers, stood firm in support of what was seen by some as a ridiculously radical building.

The political wranglings even spread into the tendering for the steelwork contract. The French companies conspired to put in a price that was 50% over budget, but suggesting an alternative, 'more sensible'

structural solution that miraculously would cost exactly what the budget was. But the Japanese steel company Nippon messed that little plan up by tendering at half the budget price, and a similar price was put in by the German company Krupp too.

Krupp got the contract but when the first gerberettes and beams were load tested they failed at half the design load. In a contract where the schedule left no slack and critics were baying in the wings waiting for the first disaster to confirm their beliefs, this would have been unnerving for experienced engineers and architects. For the young team at Beaubourg the pressure could have become too much. But even by that time Rice had learned to be confident in his ability and his engineering competence. To him it was clear that there must have been a problem somewhere along the line, and it was simply a matter of locating that problem.

It turned out that the German manufacturers had referred to the wrong code of practice in looking at the specification. The design of the gerberettes was based on relatively new structural techniques in fracture mechanics that had been pioneered in the UK for use in the design of North Sea oil platforms. Hence a British code had been referred to. The German manufacturers had used what they believed to be an equivalent, or superior

specification, from a German code of practice. Once Krupps had been convinced of the problems the units already cast were reheated to strengthen them and were saved; and the new units were made to the correct specification. And Rice's confidence in new structural theory and his ability to apply it had been justified.

Another major tremor occurred when the real client for the project, President Pompidou, died in May 1974, as the project was well on its way to completion. The building was renamed Centre Pompidou, but the problem was that Pompidou's sucessor, Valery Giscard d'Estaing, was one of the vitriolic political opponents of the project. He commanded that the top two storeys should be removed. Rice was summoned from holiday to rescue the situation. He clearly had little time for these new sceptics who had appeared on the scene and dismissed their instructions as impossible. His comments to them could be paraphrased as 'the project has gone too far and anyway if you take the top two storeys off the building will fall down'. He had a great deal of time for young engineers and architects who wanted to learn. He had no time at all for narrow-minded sceptics.

Rice went on in later projects to exploit the idea of layering of elements and systems that he had seen work well in Beaubourg; notably at La Villette, where the quality of

Fig. 4.11. Beaubourg: a transparent machine, a factory for artefacts.

Photo: André Brown

became keen to escape from the limitations of the Anglo-Saxon definitions of building design. Renzo Piano was likewise driven to find an alternative way of working. The two decided to form a practice that might give the framework for this new kind of working, called Atelier Piano and Rice. Rice remained with Arups during this experimental partnership, and no doubt some at Arups had concerns about the practical implications of the partnership. Later, Rice established, with Martin Francis and Ian Ritchie, another, more successful, experimental practice, RFR. The Peter Rice version of these experimental partnerships was that they were vehicles for trying out new forms of collaboration, new ways of pushing architectural and engineering design forward.

If Sydney had been the place where the seeds of innovation and invention had been planted, then Beaubourg had been the project where they had germinated. From this time on Rice dedicated his time in his professional work to looking for innovative ways of engineering a building and novel ways of assembling a design team where that kind of innovation would be encouraged.

the Bioclimatic Façade is achieved through a softly layered approach, to accomplish the transparency that was essential to meet the design goal. It was ideas like this, the opportunity to work with young, open-minded architects and the chance to exploit his passion for working with new materials that enabled Rice to see what could be accomplished with the will and the right environment.

After Beaubourg Rice said that he

Chapter Five
France, language and RFR

France

There were several aspects to working in France that appealed to Peter Rice. One was the difference in perception of the engineer in France. This applies to those both within the industry and the general public, or perhaps, most importantly, to the client. In the UK the professional roles have become defined in such a way that they pigeonhole responsibility and constrain cross-fertilisation. In France the engineer's role is seen more as someone who is a participant in the design process. Engineers are seen to bring to the design process particular skills that can interact with the skills of the other participants in that process.

Rice liked the fact that he would sometimes be referred to not as an engineer but as a *'géomètre'*, a word that suggests a broader design role than that typically assigned to the engineer in the UK. He was passionate about the fact that in the UK an important element in the creative melting pot was missing by having the engineers' role so tied down and constrained. He referred to people such as Jean Prouvé in France and Nervi in Italy as examples of engineer–designers whose freedom to work more broadly fostered a creative design environment that is difficult to replicate in the UK (or the Anglo-Saxon design world, as Rice referred to it).

Peter Rice admired the work of the French engineer Jean Prouvé. The fact that he dedicated a whole chapter to Prouvé in his book is testament to that.[42] And it is here, in his description of Jean Prouvé's essential qualities, that we can find what kind of engineer Rice admired and, by implication, strove to be

'He was an individual. He invented. He expressed himself. He liked to know how things worked, how they were made, how the materials that were in them could be best deployed, and to use that as the basis of his art.'

Despite this view held by Rice, it is still the commonly held view that architects are seen (and see themselves) as artists whereas the engineer does not. Rice was critical of the engineer who responded with stereotyped conventional answers to questions about an appropriate design solution. He felt that in the UK too much time and effort was wasted protecting professional positions: creating bastions that made facilitation and interaction difficult. He was very conscious of the potential to slide into entrenched positions and, in doing so, to stop being inventive. He was acutely aware that the industry and society generally tend to put constraints on what can and should be done.

These views help to explain Rice's determination to experiment with new materials and novel structural forms. It is all part of the process of challenging convention and pushing at the boundaries that constrain the potential of the engineer to make important and unique contributions to the design process.[43] But this is the view of Rice through the UK engineer's eyes. Despite his admiration of the French system of working and many other aspects of their culture certain aspects of his Anglo-Saxon heritage still prevailed.

Many of his French colleagues at RFR found him to be a baffling enigma in some ways. The French architects and engineers would have neatly coiffured heads of hair and be dressed in sharp suits. Peter Rice would be attired in a jumper with holes in the sleeve and a colour scheme taken from a packet of fruit gums. The French desks would be arranged with the contents on view sorted into small co-ordinated piles. The Rice equivalent would be a volcano of books, papers, models, photographs, faxes, computer print out and various general items of (apparent) detritus.

Yet what the volcano produced was a rich and nourishing stream of information that his co-workers would feed from. The thinking and creativity that went with it was clear, organised and well formed. It is perhaps easy to appreciate why he had a gift for mathematics, and a parallel passion for chaos theory.

Photo: Ove Arup and Partners

Fig. 5.1 Paris Charles de Gaulle: articulation and layering, an architectural language.

RFR

The work at La Villette in the early 1980s signalled the end of Atelier Piano and Rice as a formal partnership. The practical problems of trying to maintain that partnership had proven too difficult. But the La Villette project offered a fresh environment in which to create a new innovative partnership, so in 1981, as Atelier Piano and Rice died, RFR was born. Ian Ritchie had worked with Peter Rice at Arups between 1978 and 1981, on projects including Shelterspan; a system for lightweight, temporary building. He remembers the creation of RFR like this

'Peter was approached by Adrien Fainsilber (who had won the La Villette competition) to help with the engineering of the project. Adrien Fainsilber already had the engineer Constantinidis in his team, but the French government representatives in charge of the project had suggested to Fainsilber that Peter should be involved. They knew Peter from the Centre Pompidou project.

Peter asked me if I would be interested in working with him on the project… In discussion, he asked about glass design and I mentioned that Martin Francis and I had worked together on the glass wall of the Willis Faber Ipswich building while we had both been at Foster Associates.'

Peter subsequently spoke to Martin Francis, and invited him to collaborate on La Villette … 'We all met in Paris, and over dinner Peter proposed that we create Rice Francis Ritchie, and each take a one third share.'

Out of the original group of three who formed RFR it was Peter Rice who retained the significant formal long-term involvement. Ian Ritchie withdrew from RFR in 1988 to return to London to work on his own architecture. Martin Francis also moved from Paris to Antibes in the south of France to work on one of his prime passions, naval architecture. He retained an interest in the work of RFR as a potential collaborator and as a shareholder. When Peter Rice became ill Martin Francis returned to RFR to deal with projects that Peter Rice was unable to continue with.

During Peter Rice's involvement with RFR the links with Arups were strong, so strong that it was impossible to say where the boundaries were on some projects. But without Rice the RFR link to Arups has become less strong and more traditional, with Arups in a conventional engineering role on projects such as bridge designs.

As with Martin Francis, the detachment of Ian Ritchie from day-to-day involvement with RFR did not mean that Rice and Ritchie no longer wished to work together. On the contrary, Rice noted that the La Villette project showed that '*the chemistry was working*' and into the early 1990s Rice and Ritchie worked together on projects such as the Cultural Centre in Albert, France. Reflecting on Peter Rice's qualities in 1994, Ian Ritchie drew attention to '*his very genuine humanity and concern for quality and sensuality*'. Like other architects Ian Ritchie had found someone who was not only a gifted engineer, but both a friend and an ally who would share an architectural vision. That friendship and support was clearly something that both Ritchie and Rice valued.

Looking at RFR now it is clear that as Director Peter Rice sent the practice off in a trajectory that it continues along today. That is not to say that the practice has become set in a late 1980s philosophical aspic. The ethos of Peter Rice and his work through RFR was all about new challenges, new ideas or new variants on old ideas. But why did he embrace the idea of working in France with such energy and zeal? The answer to that question says a lot about Peter Rice the engineer and Peter Rice the man.

True enough, the Beaubourg project was a great joy to work on for Rice. He described the collaboration with Piano and Rogers with great enthusiasm and passion; and the international critical acclaim that the project drew was clearly a wonderful foundation on which to build the RFR practice. Added to that he repeatedly described himself as being a European with a pride and conviction that was admirable. But he had Arups in London and it must have been a difficult task balancing the demands of London and Paris. So what was it about designing in France that so appealed to Rice?

The differences in the design philosophies and practice in the UK and France are the key here. Take the pragmatic first. In France the engineer holds the main decision-making role in most building projects. Bernard Vaudeville who worked with Peter Rice on the Nuages at La Défense,

and is still at RFR, is clear about the advantages of being the engineer in the French design team. He says that it is still the case that architects '*only get to make the main design decisions for major public monuments*'. For everything else the engineer holds the trump card. That is why RFR continues to work as a multi-disciplinary team of architects and engineers. Indeed, although Vaudeville would describe himself as an architect he has also qualified as an engineer. It is not surprising then, that he describes projects where RFR are working as architects in an engineering led design team as, almost inevitably, a frustrating activity.

So, there is the first ingredient. A culture, which regards engineers with great respect and gives the engineer a title that carries the kind of caché that in the UK is afforded to medical professions.

Language

But to leave the argument there would be much too simple. There is something else about the nature of design in France, which appealed to Rice's way of working. Put simply, in France the words are used to convey the essential qualities of the design, whereas in the UK it is the drawings that the designer uses to convey the important design intentions. That view is reflected by French architect, Alain Sarfarti who, in describing his work, said that he thought that the words describing a project, rather than the drawings, should carry all the legal weight in cases of dispute.[44] If the building did not achieve the qualities described in writing then, he believed, the other collaborators in the design project such as the contractor had failed to meet the design requirements, even if the building looked like the drawings.

We can see that way of thinking in other French architects. Take Le Corbusier's description of his architectural beliefs in his article called *Precisions*. It is the words that are again the most important in describing what these new design ideas are about. Le Corbusier pointed to three areas that should be a focus of attention under the headings '*économique, sociologique et technique*'. Below these headings, again, it is words, not drawings that are used to describe the design intentions. But it is worth noting that two words appearing at the top of the list of objectives to achieve for Le Corbusier, 'standardisation' and 'industrialisation', would appear at the top of a different list for Peter Rice. For him they would be top of the list of post-industrial, tyrannical phenomena to break free from.

But the point is that in France it is words and language that hold the key to conveying design intention. Those who worked with Rice often referred to the fact that he was at his best when he had a verbal description of the design intention. Rice

freely admitted that he was '*obsessed*'[45] by the part that language played in the design process. The artist and designer Frank Stella, for instance, quotes an example where the drawings were failing to communicate what was wanted. But once Rice was fed with the information that the designer thought of the object to be designed as a twisted leaf he was away and the engineering could be invented around the core design idea. From this point the drawings became, primarily, a reference point to give an indication of location and size.

As a closing point it is interesting to note that in their collaboration after Pompidou the language which Piano and Rice chose to communicate in was French. They both appear to have found a richness and freedom by working in French rather than their native languages. The Paris based architect, Paul Andreu, described his collaboration with Rice as a ...

'...*searching together for the reason of a project, following together paths...into a world of ideas, forms and materials...[until they become inseparable from]...the project that we will develop, understand, define and finally build.*'[46]

Andreu also reports that their discussions were conducted in French, and the language itself became a vehicle for adding new ideas and new ways of

understanding the project. Talking of their discussions in French, Andreu said of Rice that...

'...*sometimes in finding his words, he alters them. And their imperfections give them a new and important meaning. He speaks of 'legerite' and, in a flash, it is no longer a question of kilos or tonnes, nor of practicality, but of grace and suspended movement.*'

Chapter Six
La Villette

The glass boxes at La Villette are, in many ways, exemplars through which Rice was, in collaboration with others, able to develop a set of ideas of ways of working in glass and steel that would be reapplied, reworked and refined in later projects. The project dates from 1981, some ten years after Beaubourg. In Rice's eyes this was a welcome return to Paris to design in a social and professional culture that he found so stimulating. It was the project that was the catalyst that enabled RFR to be formed in 1981 as Martin Francis and Ian Ritchie came together to join Peter Rice in this French experimental practice: an experiment that was to produce both fine architecture and exciting interdisciplinary collaborations.

Up until the Villette project Rice's only other major project on French soil had been the Fleetguard factory at Quimper in 1978, undertaken with his Beaubourg friend and ally Richard Rogers. That project allowed Rice to play with two of his favourite toys: tension and instability. But it was the wrong kind of project in the wrong location, perhaps at the wrong time, to be the basis for establishing a new French design 'laboratory'.

But La Villette was exactly the right kind of project. Based in Paris, the project in the Park for Science and Industry was ideally suited to a design team whose central driving force was innovation and invention.[47] It was a vehicle for illustrating that computational

techniques could be used to investigate structures in creative ways. The computer could be yet another tool to get away from the shackles of conformity. At La Villette the structure satisfied Rice's intrigue by changing its topology, changing its mode of action as the applied loads changed. In the days before the computer this would have been tedious. But Rice found the computer liberating. 'What if?' questions could be answered relatively easily. Structures could be played with to produce forms that were not blatantly simplistic.

Descriptions of the Serres (or in formal terms Les Serres et Toiture Accueil) at La Villette also reveal the importance of language that Rice was so keen to explore. Adrien Fainsilber, the French designer of the Serres, described them as a '*transparent façade*', or a '*Bioclimatic Façade*'. It was this intention, expressed in words, not drawings, that guided Rice and his co-workers in developing the technology to make the idea work. The dimensions were relatively unimportant. If the engineering could achieve a transparent façade then that was the crucial test of success.

The Pilkington structural glass system was around at the time that the design was conceived, but, in true Rice fashion, simply accepting what was already in use would have represented an abrogation of the engineer's role. Rice, instead, saw this as a key

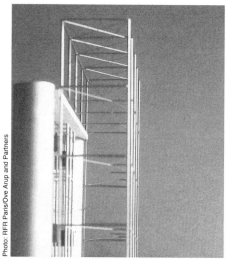

Photo: RFR Paris/Ove Arup and Partners

Fig. 6.1. Conceptual model for Les Serres at La Villette.

Photo: André Brown

Fig. 6.2. Les Serres as constructed.

Photo: André Brown

Fig. 6.3. Façade Bioclimatique.

project in which the nature of hung glass systems could be revisited and the potential for refinement and development explored. The techniques and ideas developed through the Serres project and other projects around that time led to the book written by Rice in collaboration with Hugh Dutton.[48]

Ian Ritchie recalls a conversation with Martin Francis and Peter Rice in Paris at which Ritchie vented his frustration at not being able to replace the glass fin supports, by a thin cable braced structure in the large glazed end walls of the Sainsbury Centre for the Visual Arts. The glass fins in the conventional (at that time) structural glazed wall impaired the view out rather than framing it. Some kind of cable braced structure had the potential to offer a more satisfying solution; one that gave a better idea of transparency. Rice took the coffee bar sketches away and after nearly two years of refinement, design and analysis a technique was devised that was to become the parent of many other structural glazed walls: the horizontal truss and tension net system for Les Serres at La Villette.

The panes of glass in Les Serres are connected to the light, prestressed, but apparently unstable, transom trusses. It is the panes of glass, acting as a rigid plane, that give the transom trusses their lateral stability. There are two potential axes of rotation, one along the centre line of the horizontal transom trusses, and one at the horizontal line defined by the line of connections to the glass. The conjunction of the two axes gives the truss its lateral stability.

The idea of potential instability, and resisting that potential by unusual means was part of the Rice repertoire that he used in a number of projects. It was one of the ways in which he exploited the theme of mystery or intrigue. It allowed him to shun routine solutions and add a further layer of intrigue to his buildings.

The project

On the northern fringes of Paris, La Villette was constructed as a 'City for Science and Industry' on the site of a former abattoir in an area of undistinguished markets and small businesses. The main building is a modern Museum of Technology, situated at the southern (city) edge of the Parc La Villette. In addition there are a variety of other structures around the Parc in which art and technology meet in different ways. It is in many ways a contemporary equivalent to the Paris exhibitions of the mid to late 1800s which produced the Galerie de Machines (there was more than one but the most recent is the most famous). The Paris exhibitions were meant to show how the newest ideas and newest materials could be pushed to the limit to produce impressive new structural forms. But art and grace were also important features of these new iron, steel and glass forms. The exhibitions were a parallel to, or in competition with, depending on your point of view, the London exhibitions in the same period.

In addition to the main Museum of Technology, there are structures around the Parc which took a variety of approaches to the juxtaposition of ideas of art and current technology. It says a lot that Rice was involved in more than one project at La Villette: the engineer who, by now, had the deserved reputation for regarding art and technology in building as synonymous. The technology was deficient if it did not help the artistic intentions to be realised.

A comparison of projects at La Villette in which Rice was involved is revealing. It shows that he could work in different ways with different architects. It shows that he was adaptable and that it was not his aim to modify an architectural idea in order to make it easier to analyse. It was about making real, strong and convincing architectural ideas and about making the technology add value and content.

In the Bioclimatic Façade to the museum with architect Adrien Fainsilber this is achieved in one way. In the long sculptural walkway with architect Bernard Tschumi this is accomplished with a different approach. It is the underlying philosophy of making the architectural idea work that remains constant. That and the competence and confidence to make sure it does.

The Bioclimatic Façade: Les Serres

The term 'Bioclimatic Façade' is a phrase which captures the essence of the concept which Fainsilber had for the face to the Museum of Technology that looked inwards to the Parc. It was the interface between the technology on the inside and the Parc on the outside.

It does this by framing the walkways out into the Parc with large glass boxes that contain tropical plants that are automatically sprayed with a fine mist of water at regular intervals. Apart from being practical the mists give the boxes a serene and magical quality. In fact the English word 'boxes' does not nearly do the structures justice. They deny the cool style that the Fainsilber–Rice partnership achieved here. Conservatories is an alternative way of describing the structures but the French word 'Serres' does a better job so is preferred. Conservatory is tea and scones; Serres is red wine and baguette.

There are actually three Serres: each is 30 m by 30 m in elevation and 8 m deep, back to the face of the steel museum

Photo: Ove Arup and Partners

structure. Sitting as they do against the face of the museum they play the lead role in defining the technological quality that was important for a building intended as homage to technology.

So, to allow that technology to be explicit the Serres had to be very transparent. That quality was also important in making the structures read as an interface or transition zone between inside and out. Equally, it was important that the revealed technology should be sophisticated and refined. Added to that, the Peter Rice interest in articulating the structure, so that it can be seen to be in different layers and read at different levels, would be brought into play if possible. But in essence what was required here was a glass structure that was held in place by a technological thread.

Rice's earlier IBM Travelling Exhibition structure gained transparency through the use of polycarbonate. That was an interesting experiment, but the problems of dealing with polycarbonate as a structural material meant that work on developing associated details and systems was relatively short lived. This was not so with the glass and steel structure at La Villette. The glass and supporting tension assisted steel structure developed for Les Serres was to see a series of constant refinements, variations and modifications

Fig. 6.4. Interior view showing the mullions and transom trusses.

Chapter Six
La Villette

through a variety of ensuing projects for the remainder of Rice's working life.

Structural glass had been important in bringing Rice, Francis and Ritchie together to form RFR. The Willis Faber Dumas building, where Martin Francis had worked on the structural glass fin support system for the curving glass facade, proved to be a basis for the initial collaboration of the group. The Willis Faber Dumas building had made interesting inroads into the development of glass as a load-carrying material, but at La Villette it was time to try something new.

Glass is fragile. It fails rapidly and explosively, unlike the polycarbonate of the IBM Travelling Pavilion, which is both tough and durable. The two materials have almost polar opposite properties.

One of the keys to making the architectural idea work for Les Serres was to support the glass in a minimal and unobtrusive way. At Willis Faber Dumas the trick had been to use glass fins at right angles to the glass in the external face. These fins acted as vertical cantilevers extending up from the top of the floor slab, or down from the underside of the slab. The connection was made by bolting through steel 'patches'. These patches, though small, could be read as elements on the outside face. For the Serres the connection had to be developed in a way

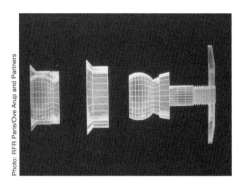

Photo: RFR Paris/Ove Arup and Partners

Fig. 6.5. Computer model of purpose designed bolt with spherical bearing.

Fig. 6.6. Glass bolting connector attached to spring to cushion against shock loading.

Photo: RFR Paris/Ove Arup and Partners

that recognised the architectural intention of maximum transparency, while recognising also the particular properties of glass under load.

Unlike polycarbonate, glass is effectively stable under environmental temperature changes. But the brittleness dictates the need for a designed solution. Sudden impacts and violent changes in load need to be avoided.

In order to spread the loads into the glass at the supports evenly over a reasonable area the patch type of fitting had been used at Willis Faber Dumas. Doing this meant that loads were not concentrated in one location. Stress concentrations lead to cracking and it is the phenomenon of crack propagation which is the Achilles heel of glass as a load-bearing material.

But the patch was intrusive as a support solution. If the glass was to be supported at a minimal number of discrete points without a plate to spread the load and was, instead, to be supported on a bearing, that bearing needed to make as much surface area contact with the glass as possible. This was accomplished by making the bearings spherical. The spherical shape gives both a large bearing area and the freedom for some displacement and rotation without causing undue stress increase at the support.

The major glass panels themselves are composed from 16 square sheets or sub-panels, each 2 m by 2 m. Because of the nature of glass, where small cracks can get propagated, leading to overall failure of the panel, Rice took the precaution of mounting the major panels on a system of springs. The effect of these springs is that if one sub-panel fails it does not transmit a shock load to the other panels, which would cause progressive failure in them too. In the event this proved wise. One panel did fail early in the structure's life, but damage was limited to the area of the fractured panel.

Apart from the supporting columns the structural steel system is principally in the form of light prestressed steel cable (and light strut) trusses. The prestressing is needed to keep the trusses light in appearance. Since tension dominates, many of the elements can be made as thin steel cables. Being prestressed the steel cables can resist the forces that would cause compression; forces normally carried by rigid elements of larger cross-section.

The prestressed trusses supporting the wall lie with their axes horizontal. As the wind load changes the pressure on any face of the Serres can change from positive to negative, or vice versa. Under extreme loadings the compression force in the cable can be sufficient to overcome the pretension. In such conditions the cable takes no load and the form of the structure suddenly changes. Some types of analysis can be confused by such phenomena. Rice's team tried a Linear Analysis approach in their computations but in a computer program which assumes simple linear behaviour a slender element cannot carry compression and effectively disappears in such circumstances. In practice at this switchover point an element running off in a different direction from the same point immediately snaps into action and the structure takes a new form to carry the loads. But an element does disappear from the calculation for an instant and this can throw the computational analysis into disarray.

Fig. 6.7. Rice sketches showing how the major horizontal trusses work.

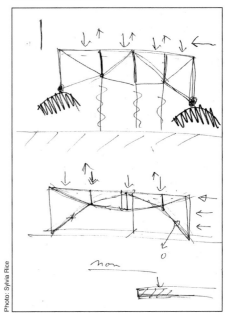

Photo: Sylvia Rice

Chapter Six
La Villette

Photo: RFR Paris/Ove Arup and Partners

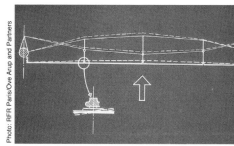

Photo: RFR Paris/Ove Arup and Partners

Fig. 6.9. The connection to the plane of glazing that resists lateral instability.

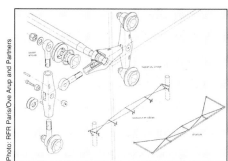

Photo: RFR Paris/Ove Arup and Partners

Fig. 6.10. The layers of articulation within the tension truss systems.

Fig. 6.8. The tension trusses: stability perched on a knife edge.

This was a point of concern to the building authorities in Paris,[49] but eventually the argument over whether the prestressed tension system would really work was won. As at Beaubourg Rice's unanswerable logic had repelled the tide of mindless application of code and rule. And the result is one that was clearly satisfying for Peter Rice. The lightness and transparency is accomplished but the technology does not completely fade into the background. It is visible, and the articulating layers are clear: glass, connector, tension truss, anchorage and column.

The tension trusses sit some distance behind the plane of the glass, and the connections to the glass are so light that they seem almost not to touch the glass. This fact, and the lightness of the tension supporting structure, enhance the feeling of transparency which Fainsilber was so keen to achieve. The resulting structure is light and almost ephemeral: the boundary between inside and out is sensitively and lightly defined. Again, these qualities are what Fainsilber was aiming to achieve in his design intention to produce a Bioclimatic Façade. As with all of Rice's important projects it was this ability to turn an architectural intention into a realised physical solution that made him such a powerful and important ally in an architect–engineer collaboration.

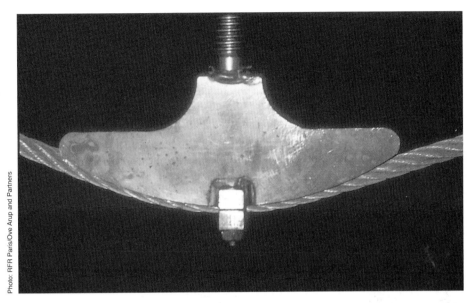

Photo: RFR Paris/Ove Arup and Partners

Fig. 6.11. Cable stay detail: typically cared for and tactile.

What we also see developing here is the idea of layering. Developed and refined at later projects such as the inverted pyramids at the Louvre, the station at Paris, Charles de Gaulle and the museum at Luxembourg, this was an idea that appealed to Rice. To him the layers and resulting articulation helped to define a clear architectural quality. To him the crudeness of one element simply landing on top of, or passing through, another meant that the whole structure became an unsophisticated morass where the lack of clarity in the articulation suggested a lack of care for the component elements.

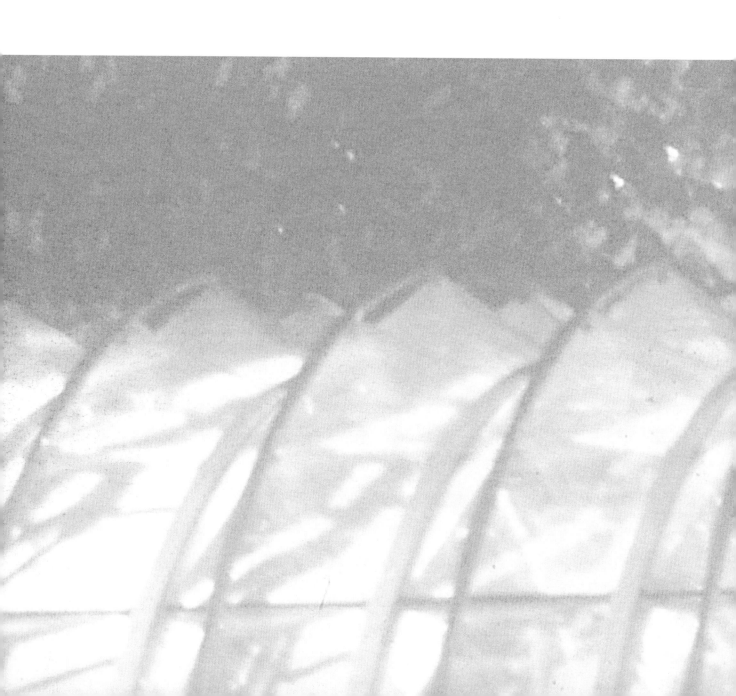

Chapter Seven
Piano and Rice

Chapter Seven
Piano and Rice

Beaubourg was a project that brought Rice, Rogers and Piano together as professional colleagues. They left the project as kindred spirits. In many later schemes Rice would work with each of his Beaubourg collaborators. Each of them had recognised the considerable engineering talent that was evident in Rice's contribution. They would want to exploit that talent again in future schemes. The fact that Beaubourg led to lifelong friendships too was a bonus.

In Renzo Piano's case it is difficult to identify whose influence was strongest in developing the many mutual ideas and motivations that Rice and Piano shared. They both talked of the importance of experimenting with new materials, about a human scale, about leaving the mark of the craftsman in a building, and many other matters. Take the words from Piano's mouth and put them into Rice's and, apart from the change in accent, it would have been impossible to tell who was speaking. They both liked the term 'piece' to describe a crafted element in a building, they both sought ways of articulating elements. But the argument about who influenced whom is an argument not worth having. In regular meetings they developed common ideas and ways of working, and that is what is important; the fact that ideas could be shared, traded and refined in an atmosphere that allowed engineering and architecture to thrive in equal measure.

The result was not dogma, but a broad philosophy; an approach that was guided by a range of potential, but not imperative, goals. This is a comment that came from Renzo Piano, but could just as easily have come from Peter Rice.

'I can hardly see a separation between shape, function, structure, technology, technical equipment and science; between science and art there cannot be a barrier; they speak the same language and require the same energy.'

Their common drive, common interests and close friendship led them to set up a formal Piano–Rice partnership that they envisaged would tackle projects involving architecture and engineering. Their approach, they intended, would be freer and less constrained by professional territories than in conventional projects. But much of the work that the formal partnership of Piano and Rice managed to bring in was divorced from the world of engineering. On top of that trying to work across international boundaries proved to be difficult even within Europe.

The result was that most of the work that came in was either abortive or undertaken by Piano alone. The exception was the project to design a prototype car for Fiat where the aim was to design a car for the 90s. What is particularly interesting here is the comparison between the building and car design industries at that time. Buildings (Rice's buildings at least) were designed as individual objects whereas cars were being designed by modifying and remodifying existing prototypes.

The result of the collaboration was that a new design for the body of the car was developed but on the whole the experimental partnership with Piano proved frustrating for Rice. A combination of inappropriate projects and differences in Pan-European working practices proved too limiting. The formal partnership between them ended around 1980. It is no coincidence that as Piano and Rice as an experimental partnership died the new practice of RFR began to form.

But these events did not mean that the collaboration between Piano and Rice had ceased. The reverse is true, in fact. Piano, through the Renzo Piano Building Workshop, and Rice, through a combination of Arups and RFR, took on several important projects after the demise of Atelier Piano and Rice. And whenever their mutually supportive talents came together the result was, almost without fail, a building that brought together quality and innovation.

In this context three buildings are described here: the IBM Travelling Pavilion (1981): the de Menil Gallery (1981): and Bari Football Stadium (1986).

IBM Travelling Pavilion

In the same year as the La Villette Bioclimatic Façade project was undertaken Rice and Piano worked on a travelling exhibition building for the IBM Computer Company and the Menil Collection. On the face of it these do not have much in common but there are some interesting comparisons to be made. Intellectually and professionally they have much that they share.

In the case of IBM and de Menil the prime reason for this is that it is the same pairing of Piano and Rice. Together they had developed a particular approach and guiding principles. But they were just that, principles, not dogma or fixed ideas, but principles. This means that the viewer needs to appreciate the underlying philosophy rather than the superficial appearance to see the strong family resemblance in a Piano–Rice collaboration.

Photo: Ove Arup and Partners/Harry Snowden

Fig. 7.2. The exhibition building, May 1984.

Fig. 7.3. Polycarbonate and laminated timber.

Photo: Ove Arup and Partners/Harry Snowden

Fig. 7.1. Piano's model of the IBM Travelling Exhibition building.

Photo: André Brown

The reason that Rice gave for the choice of polycarbonate as a principal material in the IBM building was that it was clearer than glass, letting in a better quality of light, but without the colour bias that glass exhibits. A particular appeal to Renzo Piano was the potential of the 'crystalline form' that the building would take on by having an outer skin of facetted polycarbonate elements. But other reasons must have had a major influence too. As a travelling exhibition the fact that the material is lighter and less brittle than glass would have clearly made it easier to handle and transport. The other materials, too, are chosen with those qualities partly in mind. The skeleton that works in conjunction with the polycarbonate uses materials that are relatively light, but with a good strength to weight ratio. They are timber for the frame and cast aluminium for the joints.

Fig. 7.4. Special axle connection between the polycarbonate and laminated timber frame.

Photo: André Brown

Chapter Seven
Piano and Rice

But the factor that was just as important was that the client was a good one, being at that time a company perceived as being at the leading edge of contemporary technology. As a consequence they were sympathetic to the Rice and Piano commitment to experimenting with new materials and new technology.

IBM wanted to reinforce their market position by taking a travelling show from city to city in Europe, staying in each for three to four weeks and spreading the word that computers were the future and that computers meant IBM, also known at that time as 'big blue'. On this occasion they wanted to be 'small green'. The concept given to Piano and Rice was the pavilion in the garden: there was an emphasis in linking the inside and the outside.

Now for other engineers the mixture of polycarbonate, timber and cast aluminium, and the requirement of rapid demountability, would have been more than ample ingredients with which to work. But Rice saw a further opportunity. Polycarbonate is often used as a non-structural material, but if it could perform a structural function then there would be the possibility to lighten the rest of the structure and to do away with some other elements.

But polycarbonate has a number of rogue qualities that require non-standard solutions. These solutions do give the building a particular quality at the detail level. Some would argue that the complexity of the solution that had to be adopted represented a degree of overkill that detracted from the potential clean simplicity that might have been achieved if an alternative approach had been taken. For Rice the answer is clear. The complexity helps define the design but, as always, there is a logic and reason underlying the complexity. It is not a complexity simply for the sake of it. It is a complexity that is born out of providing a solution to an engineering problem. And it is a complexity that sits against and contrasts with the clean, clear simplicity elsewhere in the building.

Rice, typically, met the various problems head on, knowing that there would be a solution. The principal problems with polycarbonate were these: it displays considerable thermal movement and this expansion and contraction has to be accommodated; it has a poor response to heat, losing strength appreciably when the temperatures are over 70 °C. Although it is a tough material it additionally exhibits cracking under prolonged stress. So on the one hand Rice was working with glass at La Villette, a material which is strong but fragile, while on the other he had to concern himself with polycarbonate for the IBM building: a material with properties which are the opposite of glass. It is tough but weak.

How, then, can the structural function for the polycarbonate elements be provided while allowing a large degree of thermal movement, and protection from continuous loading? The answer is revealed in the joint where the polycarbonate meets the laminated beech rib. The polycarbonate is held at a distance of about 120 mm from the timber (Fig. 7.6). To maintain structural integrity, while providing freedom for thermal expansion, a rubber block was inserted in the joint between the polycarbonate and the struts. At the polycarbonate end of each of the struts an axle was placed which allowed movement in one direction only. Rigidity was maintained at right angles to the axle. This joint and its connection to the timber gave a character, a personal fingerprint, to identify the building.

The way that the cast aluminium node looked and felt appealed to Piano, but again there was also a structural job to do that would, to some extent, define its form. It was connected to the timber rib. Timber is strong in tension, but more limited in shear so the disposition of the joint was configured so that the shear load could be accommodated (Fig. 7.5).

Again Rice plays up the solution and the difficulties were played down. In fact, although the jointing detail is very clever, on its own it did not give the whole answer to the problem. On a much more pragmatic

Photo: Andre Brown

Fig. 7.5. Aluminium node.

Photo: Ove Arup and Partners

Fig. 7.7. The components on site.

Fig. 7.8. View from inside.

Photo: Ove Arup and Partners/Harry Snowden

Photo: Ove Arup and Partners

Fig. 7.6. Axle connection in the building as constructed.

Chapter Seven
Piano and Rice

level the engineering design team were able to argue that, because of the temporary nature of the building and its construction, design imposed/wind loads could be reduced significantly. Reducing the design load clearly meant that dealing with the structural shortcomings of polycarbonate were eased significantly.

The other aspect that is worth noting is that there were actually two IBM Travelling Exhibition stands, not one. In order to get over the problem of rapid demountability and remounting two kits of parts were made and these leapfrogged each other around the country. So, as soon as the exhibition was finished in one place the building was ready and waiting to receive it in the next destination.

But the quality of the final building was the real testament to designers who cared deeply about the designed product. Rice and Piano shared common goals and were each willing to let the other's skills influence the nature of what they were to produce together.

Like the Beaubourg gerberettes and the Menil leaves the special connections on the IBM building gave a particular characteristic mark, like the marks left by carpenters on early timber structures. They reveal a concern for craft and architectural quality.

The Menil Collection

The Menil Collection is a project that Peter Rice found particularly satisfying. Apart from the fact that the product is a lovely and successful building there were aspects to its production that gave opportunities for experimentation.

The first was the possibility to try out unconventional materials and, in particular, find new applications and appropriate forms for them. The second was to work with a gifted architect in an unconventional way;

Fig. 7.9. The pavilion under construction.

Photo: Ove Arup and Partners

Fig. 7.10. The building in use.

Photo: Ove Arup and Partners

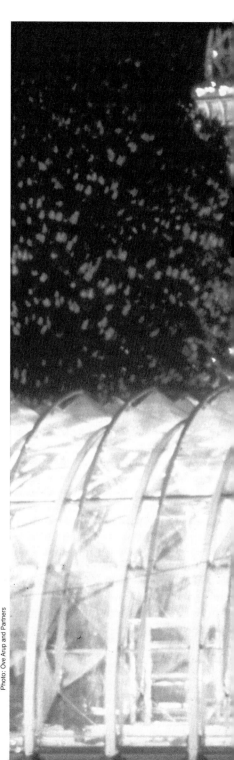

Photo: Ove Arup and Partners

Fig. 7.11. IBM exhibition building at night in Paris.

breaking free from the constraining rules of dated working practices. Third, but no less important, was that in collaborating with Piano he was working with a friend: with someone who also wanted to break down the barriers created by conventional practice and with someone who valued craft in a building and was a superb craftsman himself.

In 1981 Dominique de Menil commissioned a new building to house her very prestigious collection of symbolist and primitive African art. The qualities that she sought were a relaxed relationship between the viewer and artwork, achieved through a non-monumental space with a domestic scale. The Menil Gallery is in Houston, Texas, and light was to be a particular determinant of the architecture.[50] In particular the lighting of the exhibits needed special attention and a carefully worked architectural solution.

The client, Mrs de Menil, had bought all of the works of art to be displayed herself. She was, understandably, committed to producing a building of originality and quality: a work of art to display works of art. In particular it was proposed that natural light should be taken into the gallery, and not shunned, or played down, as in many other display galleries. From this point on light became the key driving force.

In addition to having Piano's desire to produce crafted elements, a desire that Rice shared anyway, there was now a design intention that Rice could use as the basis for his decision making. The design was now about crafting roof elements that would throw a good quality of diffuse light into the gallery for the artworks to be viewed at their best. As in other projects having a design intention like this to guide him Rice was able to manipulate the technology to realise the intention. In doing this he recognised the importance of Tom Barker's contribution, who worked alongside him at the Arup end of the collaboration.

The effect of light on exhibits is cumulative over time, and what dictates the level of deterioration is the total amount of light that a work has experienced over its life. Consequently, the thinking was that higher levels than normally recommended could be accepted for certain periods of exposure if at other times the work of art was subject to lower levels of exposure. Having taken the decision to use higher natural light levels for display, therefore, a consequent decision was to rotate the collection, with objects going into and coming out of storage such that their total exposure to light was less than that recommended. At any one time about 200 objects are on display and this represents about 10% of the total collection.

LIGHT

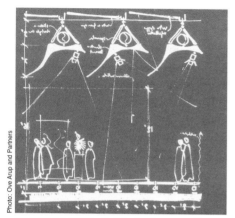

Photo: Ove Arup and Partners

Fig. 7.12. Architect's sketch showing proportions and lighting.

Fig. 7.13. Part of the completed building where the roof becomes an external element.

Photo: Ove Arup and Partners/Peter Mackinven

Fig. 7.14. Physical model at the site.

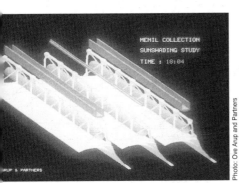

Fig. 7.15. Computer model of the roof structure.

The storage area remained accessible if works needed to be viewed as a one-off for a particular reason, but for the most part it would be the naturally lit gallery where the works would normally be seen. Rice described this as a philosophical solution to a particular architectural dilemma.

An additional device that was used was to take account of the fact that the ultraviolet end of the light spectrum is the most damaging and that this kind of light is absorbed during reflection. A system was therefore needed which could bounce the light into the gallery space off at least two surfaces.

What was required was a roof light system that acted like a sophisticated venetian blind. The fins, that ran in rows across the roof plane, became referred to as leaves. But these leaves needed, ideally, to be curved in some non-uniform way so that they bounced the light in the desired way. The actual form was to be prescribed by running extensive computational analyses that would iterate towards an ideal shape. But the leaves also needed to be sympathetic and appropriate for the very particular gallery setting. Tasteful, but reserved.

Typically, for a Rice and Piano project, a material that has had limited use in the construction industry was brought into play. This time it was ferrocement. Renzo Piano had already had experience of building boats in the material in which short strands, rather than long bars of steel, are used to give the composite material the tensile strength and ductility which concrete alone lacks. It can be made into strong, dense sheets which, because the material is cast, can be moulded into a wide range of forms. In this case it could be worked into the shape that would achieve the optimum performance in terms of reflections.

During their experimental five year partnership as a design team Rice and Piano had searched around for a project in which ferrocement might present a good solution. In addition, in 1977 Rice had undertaken a study on behalf of Pilkington Glass in collaboration with the third member of the Beaubourg team, Richard Rogers, to investigate the potential of glass fibre reinforced cement for roofing units.

Ironically, now, with the formal Piano–Rice partnership dissolved, the project came along that they had been waiting for. One in which moulded ferrocement forms offered a solution to a particular architectural problem. One that would allow the development of a composition of crafted individual forms. Those forms would suggest the work of the human hand which both Piano and Rice strove for in their design collaborations.

Photo: Ove Arup and Partners

Fig. 7.16. Sketch of the leaf.

Photo: Ove Arup and Partners

Fig. 7.17. Steel mesh being laid into the cement.

Photo: Ove Arup and Partners

Fig. 7.18. The composite ductile iron and ferrocement roof trusses.

Because the sheets of ferrocement can be as little as 2·5 to 3 cm thick, at the scale of the overall piece they convey a lightness and delicacy that is difficult, or impossible, to realise in conventionally reinforced concrete. The word 'piece' is used here deliberately. It is the word preferred by Rice for the same reason that he used the word to describe the gerberettes at Beaubourg. It suggests the artist is involved and that each element is a sculpture in its own right, in which the form and material have been carefully chosen. It suggests humanity and denies the mindless industrial process a place. But having said that Rice was rather hesitant about using the words 'art' and 'sculpture' to describe these structural elements.

Although he actively promoted the sharing of the design process with his fellow collaborators, he was also very keen to recognise that each of the participants in the process have particular skills which enrich the design, but cannot be usurped. Rice was the engineer, and wanted to remain respected and valued for being just that.

In the Menil Gallery the supporting columns for the roof structure are about 40 feet (12·2 m) apart, and the roof structure is a truss that spans between the columns. Each leaf therefore acts as the tension member in the truss.

The ferrocement leaves have the particular problem, from a structural point of view, that they cannot be adequately tested post-production. The performance characteristics therefore have to be met through quality control in the production of the element: a reliable and repeatable process is a must. But like the gerberettes in the Beaubourg project, if the quality is to be attained, then this is a price that has to be paid.

Unlike conventional reinforced concrete, for which there is a wealth of information that can be used as the basis to test for acceptable quality and performance characteristics, ferrocement is a relatively virgin structural material. Part of the engineering design process was, therefore, to define criteria for the production and quality of the ferrocement leaves.

The outcome is again one that is unconventional in structural terms. The performance characteristics are achieved through monitoring the production process, not in checking the individual pieces. This resulted in the first 24 or so ferrocement elements being rejected.

Rice's team took the unusual step of telling the factory manager that they would take their business elsewhere if he set foot on the factory floor again and advised his workers what to do. This is an example of that particular side of Rice's engineering persona that reacted badly to blind acceptance of established methods. It was

not that Rice had a problem working with contractors and material production factories. In fact Corbett Gore, who was one of the contractors at Sydney, was praised by Rice as an open-minded innovator where circumstances dictated it. The problem in the production of the leaves for Menil was with the narrow mindedness, and belief that standard methods were the only ones that worked. That was what he was being presented with and that was what he objected to.

So, with the lower, light reflecting element of the truss resolved what would be the appropriate material for the other elements of the truss? An uncommon structural material of course. This time it is ductile iron. With the leaves to be made in ferrocement in a curved form that resembles (the colour reinforcing this) a skeleton the desire was to have the remaining elements of the truss made in a moulded skeletal form. Most contemporary structural steel is hot rolled. This means that it is uniform in cross-section and, therefore, not suitable.

Ductile iron is a form of cast iron. Being cast the elements can be irregular, in this case skeletal in form. As the name suggests, unlike early cast iron which was brittle and failed explosively, ductile iron has flexibility and tensile strength, achieved through removing impurities and controlling the make-up of the material including its carbon content.

The alloying materials added to cast steel make it more viscous than cast (ductile) iron. Consequently, it flows less well into moulds so that forming intricate shapes proves difficult. Since the forms envisaged for the de Menil trusses were rather fine and irregular, ductile iron proved to be the appropriate choice. Piano and Rice agreed that the whole truss should, as far as possible, be read as a contiguous skeletal structure, with the elements flowing from one to the next. The ductile iron solution gave the additional benefit of avoiding bolted plate connections so on this count and for its finer, more detailed surface quality, it was also a winner.

The combination of the complementary ferrocement and ductile iron elements produced a piece (again the word used deliberately) which could be read as a whole as well as being read as a combination of parts. As such it helps define the architecture and the very particular quality of the gallery.

The shape of the leaf finally adopted is actually only partly governed by the need to reflect light in a particular way. It is also, Rice freely admits, partly the result of the architect's hand. There is also some fairly complicated engineering going on here too. The ferrocement leaf swells at its junction with the ductile iron truss to which it is

Chapter Seven
Piano and Rice

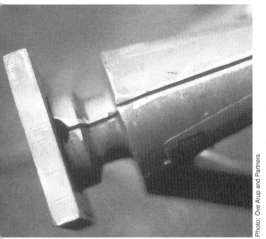

Fig. 7.19. End connector on a ductile iron piece.

Photo: Ove Arup and Partners

Photo: Ove Arup and Partners

Fig. 7.20. Ductile iron components after cleaning up.

Fig. 7.21. Bolted joint used to connect the ductile iron elements.

Photo: Ove Arup and Partners

Fig. 7.22. Final assembly of the triangulated iron elements.

attached in order to avoid stress concentrations at the points of connection. But Rice was keen to play this aspect down and to play up the architect's role. He recognised Renzo Piano's particular skill and talent in his ability to produce pleasing three-dimensional forms that worked well with the overall architectural intention. It was this kind of talent that Rice saw as the reason why architects and engineers should work together in a mutually beneficial collaborative environment. It was also the reason why he was so critical of Santiago Calatrava, who claimed to do both jobs but, in Rice's opinion, produced built objects that were, as a consequence of lone working, compromised and deficient.

In considering all the other aspects of the Menil Gallery design it is easy to take for granted certain parts of the engineering, and the engineer's role. Ductile cast iron and ferrocement are not common structural materials, but here they are acting in a composite fashion in a truss. The elements of the truss are made as flowing forms. These forms need analysing and the sizes establishing at some point. It is typical of Peter Rice that in reports of the Menil Gallery he does not consider this as a potential problem. To him this matter would be interesting rather than daunting; a reason to do it rather than a reason to not do it.

Bari Stadium

Outside engineering and architecture Peter Rice had a wide range of interests that he cared passionately about. Photography, theatre, food and art all gave him delight in different ways. He was a keen sports enthusiast too, and had a particular liking for horse racing and football. Within sport he had his preferences, though. It is not surprising that it was the supreme artistry of Italian soccer that appealed to him much more than the clinical premeditated precision of American football.

As his fame grew Rice was able to take advantage of his position by getting involved in schemes that engaged him with his

Fig. 7.23. Piano's full scale model of the Menil roof elements.

Photo: André Brown

Chapter Seven
Piano and Rice

Fig. 7.24. The gallery under construction.

Photo: Ove Arup and Partners

Fig. 7.25. The Menil Gallery.

Photo: Ove Arup and Partners

sporting passions. He had an involvement with the design of a new stand at Epsom Racecourse in collaboration with Richard Horden, but the big project that he clearly loved was the Bari football stadium, in southern Italy.

Apart from the fact that it would house football, and the home games for Bari FC, to Rice, another appealing feature of this scheme was that it was a community stadium that was intended to have a range of uses.

The project was executed in conjunction with the Renzo Piano Building Workshop, based in Genoa, for whom Ottavio di Blasi was project architect. Despite Piano and Rice's shared interests in other areas, and their personal friendship, football was something that Renzo Piano could take or leave. For Piano this was a stadium project. For Rice it was the chance to produce a building worthy of the sporting artists who would grace it and the world media stage that would see it as part of the 1990 World Cup Football Championships.

Rice put together a new team at about this time. Some months into the project Tristram Carfrae, who became his right-hand man for many subsequent Piano–Rice projects, joined him on this team. The stadium was to be for 58 000 all-seated spectators and was initially intended for football alone, but an athletics track was

Fig. 7.26. Piano's model of Bari Stadium.

Photo: André Brown

added to the design brief. Because events there were to be broadcast to the world other aspects of the design brief were that the stadium should be photogenic, make suitable provision for television cameras and provide good locations for hundreds of journalists.

The scheme was based around the idea of a concrete bowl shaped stadium, from which extended cantilevered, curved fingers to pick up the lightweight roof structure. The stadium was to have two tiers, the upper one perceived as floating above the lower one which sits in a man-made crater.[51] To maintain the floating, flying saucer analogy the underside of the upper tier was kept as clean and smooth as possible. To help accomplish this the radial segments were made in high quality precast units and the pair of supporting in situ annular beams, which ran around the whole stadium, were hidden in the depth between the soffit of the radial units and the seating above.

There is a clear band of daylight between the upper and lower levels (Fig. 7.32). In addition, the upper level is broken down into 26 distinct segments so that the upper stands read as separate articulated elements rather than as a large continuous oval structure. Wherever they sit the

Chapter Seven
Piano and Rice

supporters are never more than 190 m from the action.

The cantilevered fingers that support the roof reinforce the subdivision; the main fingers project upwards from the edges of each of the main segments. Consequently there are two bands of light (the gap between the upper and lower tier, and the light transmitted through the fabric covering) which break down the scale horizontally. The subdivision of the upper concrete tier of the stand, the curved cantilevered elements and the roof substructures break down the scale vertically. The result is that the stadium reads as a (very large) whole, but it can also be read at the level of the spectator, the subdivision and the reduction of the scale in a controlled way helping to define sub areas within the stadium that each spectator can relate to. Even coping with a crowd of 58 000 Rice and Piano remained conscious of the need for a human scale in their building.

After the tender stage the responsibility for the concrete elements was handed over to the Italian contractor. Piano and Rice concentrated on the Teflon roof which was conceived as a set of 26 'petals'; one over each section of the upper tier. The petals vary in size: 27 m wide at the centre line and 14 m at the narrowest ends closest to the pitch. The main supporting element for each petal is a pair of curved, tapering cantilevered ribs made as steel rectangular hollow sections.

Fig. 7.27. The view of the stadium approaching on foot.

Photo: Ove Arup and Partners

Fig. 7.28. During construction: cantilevered fingers and fabric cover.

Photo: Ove Arup and Partners

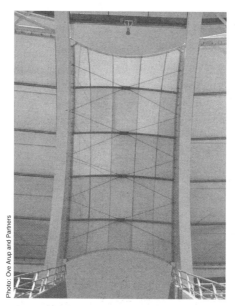

Fig. 7.29. Roof viewed from below.

Photo: Ove Arup and Partners

Fig. 7.30. Bracing above the fabric.

Photo: Ove Arup and Partners/Alistair Lenczner

A U-shaped tubular truss that serves a multiple function connects the free ends of the cantilevered ribs. Being rigidly connected to the free ends of the cantilever braces the petal structure by portal action. But the U-shape means that it can also accommodate an access walkway around the whole structure and can support the floodlighting without the need for an independent structure.

The main ribs of each of the petal structures are augmented by three chord braced tubular arches, stiffened by purlin-props that span between main ribs to form an overall petal structure. The PTFE coated glass fibre fabric that is stretched over the petal structure has a translucency of 13% which is enough to give it a very light lustre in daylight. That lustre is enhanced by the fact that the tubes, chord braces and purlin-props are all made in stainless steel to ease maintenance. The fabric is clamped to the petal structure by a series of bars threaded through the edges of the fabric. But the curvature in the fabric is relatively shallow so a bracing system of 'lazy' cables sits over the top of the membrane in the valleys to prevent curvature inversion in conditions of wind uplift.

The mixture of forms and materials at Bari produces a delightful and distinctive structure that is typical of a Piano–Rice collaboration. It is a building that can be read as an overall form or piece by piece. It bears inspection at all levels. In the evening uplighters illuminate the underside of the upper tier and the petals above in a spectacular way; one that gives the stadium the kind of attention and reverence normally reserved for a cathedral. But, after all, it is a cathedral for sport.

Photo: Ove Arup and Partners/Jack Zunz

Fig. 7.32. During an event at Bari.

Fig. 7.31. Inside the stadium just after completion.

Photo: Ove Arup and Partners

Chapter Eight
Rogers and Rice

Chapter Eight
Rogers and Rice

Lloyds of London

Lloyds was a project in which, somewhat unusually for Rice, the professional arrangement was conventional, with Rice's team at Arups acting as engineer to Rogers as architects. That suggests a stuffy formal relationship between Rice and Rogers, which was anything but the case. The two had a regular Monday meeting booked in the diary at which they could talk about potential projects and new ideas. It was a venue to keep the architecture and the engineering churning together so that the two could interact.

It takes a good engineer to appreciate the needs of the architect and to supply them with a sounding board that will help them to realise a project's full potential. But it takes a good architect to listen and invite the engineering skills to the architectural party. Rogers did that in a way that was both democratic and encouraging. These were qualities that Rice shared with Rogers. The result was a design environment that produced innovative ideas and fine architecture.

In the Lloyds project Rice was, in particular, able to try out new ways of working with the tried and tested material of concrete. Typically, he was always seeking to push established materials and find different ways of exploiting their properties. Since the

Photo: Ove Arup and Partners

Fig. 8.1. Lloyds: the philosophy of making the technology explicit.

Fig. 8.2b. The completed atrium.

material was to be exposed, and not only exposed, its bareness was to be one of the features defining the architectural quality, optimising the quality of the concrete pieces was *a priori* high on the agenda. So was articulation.

Rice[52] noted that in the Lloyds project it was to be the work of Louis Kahn that he took as a source of inspiration. In particular the Yale University Art Gallery with its cast in situ concrete floor grid was cited by Rice as an interesting precedent. Rice was always keen to point to good examples of engineering and engineers that he admired, playing their part in this way, whether it was Jean Prouvé or Louis Kahn, it was not the

type of work that they did that mattered. It was their approach to innovation; their desire to keep pushing at the frontiers of engineering in whatever way they could.

At Lloyds the high quality of the structural concrete that was required by Rogers, if he was to realise his design intentions, had to be met in a reliable and consistent way. One of the major shortcomings of cast concrete is the joint between one cast section and the next.[53] Quality control is crucial in obtaining good results at this junction. But even with good quality control there is always the problem that the wet concrete is poured into the moulding formwork. Until the concrete is hardened and the formwork is struck it is to some extent a case of crossing the fingers to hope that there has been full penetration into the mould, and no leaking at the ends. By minimising the number of these joints that were to be exposed this problem could be reduced.[54]

The means to that end were by careful and appropriate amalgamation of in situ and precast concrete. As far as possible the structure avoids having in situ concrete elements meeting each other. The precast elements thus provided the means to ensure a good quality of finish in the relatively difficult areas to cast. But they also gave the structure a degree of articulation that is rarely achieved in concrete structures.

Fig. 8.2. Cross-section through the atrium.

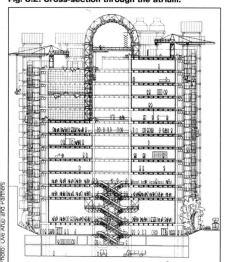

Photo: Ove Arup and Partners

Fig. 8.2c. The room-atrium space, as conceived.

Photo: Ove Arup and Partners

95

Chapter Eight
Rogers and Rice

Photo: Ove Arup and Partners

Fig. 8.3. The floor grid-main beam-column connection taking shape.

Photo: Ove Arup and Partners

Fig. 8.4. The precast bracket.

Fig. 8.3a. Plan at first floor level.

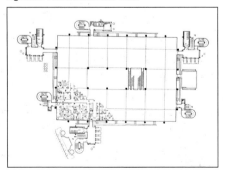

Fig. 8.3b. Main floor structure.

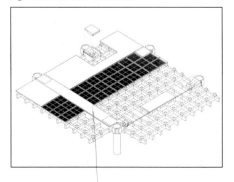

2 ribs closing coffer making u shaped beam

The building was designed to be read as '*this part does this job and that part does that one*'. The articulation through the adoption of precast joints allowed that to happen; and Rice understood this. As at Beaubourg he saw his role as engineer as being to maintain the structural integrity while allowing key aspects of the architectural idea to be realised. In this case Richard Rogers regarded the quality and articulation as key parts of the architectural game that was being played. Rice was able to maintain a rhythm of 'precast–in situ–precast–in situ', throughout the important parts of the detailing. So, in

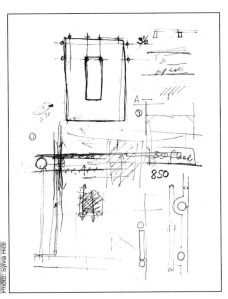

Photo: Sylvia Hide

Fig. 8.5. Rice's sketch: working through the interaction of structure, servicing and planning.

typical style, although the jointing and use of precast concrete in this way was somewhat unusual, Rice said it could be done, knew he could make it work and made sure it did work.

This was a building that attracted critical acclaim, a building where the supportive engineering allowed it to reach its architectural goals, and not least a building that allowed him to work with his long-standing friend Richard Rogers. But it was a project in which the architect and engineer played their traditional roles. For Rice part of the challenge and potential was therefore

missing. He did not want to usurp the architect's role, simply to play a part in a symbiotic, and therefore more rewarding and complete, design process. That is why he strove to experiment with non-traditional ways of working wherever possible. What Lloyds shows is that even within a conventional professional relationship, with the right framework and between consenting adults, there is still the opportunity to innovate and to challenge standard ways of building.

The building

Broadly speaking, the Lloyds building is a rectangular plan 68·4 m x 46·8 m. The dominating feature on the inside of the building is the massive open atrium in the core, which rises through the whole building from the ground floor, through 12 levels to the roof structure. At each of the 12 levels there are galleries which are constructed as rings around the perimeter of the atrium.

At ground floor level there is 'The Room', a double height space, and below this (and street level) is a public and service area with restaurant and Library. The main vertical structure is a set of 1050 mm diameter cast in situ columns. There are eight columns on the inside of the atrium and twenty outside the glazed walls on the outer edge of the building. These columns support a square 1.8 m x 1.8 m grid of concrete beams, and

Photo: Ove Arup and Partners

Fig. 8.6. Column to bracket transition: insitu-precast-insitu.

Fig. 8.7. Column and bracket: just after removal of formwork for the upper column.

Photo: Ove Arup and Partners

Rogers wanted a consistent structural solution across the whole floor, without major deviation in form at the edges or corners. But the spacing of the columns meant that the grid had to span a distance of 16·2 m and the floor depth was in danger of becoming excessive.

The solution that Rice's team came up with was to adopt a mixture of change in structural form and change from ordinary reinforced concrete to prestressed concrete in the critical locations. The main grid beams were 550 mm deep and 300 mm wide. Rice recognised that the first part of the solution was to respond to the natural flow of forces in the grid. The lines defining the column grid were, therefore, beefed up to attract load

to them by connecting two ribs together, closing the coffer, to form an inverted U-beam between the columns. To emphasise the difference in stiffnesses the remaining beams were made half depth.

With the grid defined in this way, it was then possible to pick up the grid at any point on the building face. The columns are actually spaced at 10·8 m intervals around the perimeter of the floor.

To keep the structural depth down the U-beams were prestressed. If these beams had been in plain reinforced concrete they would have been packed with reinforcement and it would have been difficult to achieve a high quality of finish on the concrete surface. Increasing the depth was another alternative, but the floor-to-floor height was already 4·5 m, of which 1·5 m was taken up by the floor itself. Prestressing also offered the additional advantage that net deflections and cracking were reduced.

However, even with these tactics adopted the out-of-balance loads at the corners of the floor slab could not be dealt with, so increased load capacity was needed. One way to deal with this would have been to increase the beam depth in the corner bays to full depth. But this potential solution raised problems. One was the practical problem that the exposed service runs would have been interfered with; a

second minor problem was that the formwork would have been more complicated. But the main problem was that it did not fit with the concept of how the building should be read. Put simply it would not have looked right. Technology would have dictated an uncomfortable solution.

The answer, in the end, was reasonably direct. In the corner bays the half beams became prestressed too, so that the whole bay became a prestressed coffered slab with prestressed edge beams. The result was a controlled and consistent appearance.

At Beaubourg Rice and Rogers had articulated the connection between the column and beam by using the cast steel gerberettes, which sat on bearing pins. The gerberettes extended inwards to provide a seat for the long span beams that traversed the main building spaces. In the Lloyds building the two sought a similar articulation. This time the 'hand' that extended out from the column to pick up the floor was in the form of a precast concrete bracket. Bracket is the term used in the descriptions of the buildings, hence its use here. But the term does the element a great disservice and belies the sophistication that this object embodies.

The bracket performs multiple functions. It hides the construction joints in the column, it receives the prefabricated bracing that provides lateral stability, it

Options: - Plain reinforced concrete
- Increasing depth.

Photo: Ove Arup and Partners

Fig. 8.8. During construction showing the bracing elements.

Photo: Ove Arup and Partners

Fig. 8.9. External view after completion: revealing articulation and subdivision.

Chapter Eight
Rogers and Rice

extends out to support the slab, and it transfers the loads from the floor structure to the column. That last function was obviously onerous. In fact, in order to make the bracket perform its structural function this way it is packed with steel (Fig. 8.4). It could not be much smaller than it is.

The form of the bracket was a hollow cylinder. To form a column–floor junction the process was this. The first length of cylindrical in situ column was cast. The precast bracket was then lowered into position at the head of the column and held in place. Concrete poured into the cylindrical void then fixed the bracket in position and made the connection to the column. Good tolerances on the concrete elements meant that there was little chance of the concrete leaking in the joint so the column–floor junction ends up being a very neat affair expressed through the high quality precast bracket.

Both the horizontal and vertical loads are transferred through purpose made bearings that sit between the U-beams and the brackets. The bearings and the brackets are all part of a typically articulated Rice system. The components of that system are the in situ column, the precast bracket, the cast in situ grid, the prestressed U-beam, the precast yolk, the steel panelled permanent formwork, and the structural topping. Each element has a part to play in the same way

that the elements of Rice's steel and glass structures are expressed. In keeping with the architectural desire to maintain a consistency and clarity internally the diagonal wind bracing elements in the main structure are confined to the external face of the frame.

Although it is true that Lloyds was originally intended to be a steel building there was a major rethink when the move to concrete came. It has been said that Lloyds is a steel building built in concrete. But that misses a crucial point. Rice says candidly that

'Our aim was to exploit the natural qualities of concrete while trying to achieve the visual articulation and legibility normally associated with steel.'[55]

In this respect Rice was consistent. Always seeking new ways of manipulating and incorporating materials; old or new. To him this was simply another opportunity to examine how the properties of a conventional material could be worked with in an unconventional or untried manner.

In the same way that it is interesting to play with steel structures, making them look like concrete structures by using large hollow sections and all-welded joints, this is a concrete building that experiments with connection and articulation that is more likely to be found in steel. But the point is that this is about experimentation and about learning. It is about adding to the pool of

engineering knowledge. Without this kind of experimentation we accept convention. Technology stagnates and the palette of creative tools is diminished.

Lloyds is a concrete building. By making it out of concrete the material has been exploited in ways that can be instructive elsewhere. Instructive, not copied. Where it leads to in the next building project is just as important as the project itself, and for Rice, later projects would be enriched and informed by the lessons learned at Lloyds.

This essential desire to exploit the properties of materials was a driving force behind much of what Rice did. But there is more than that. He was critical of engineers who eschewed the potential of clarity and discernment that could be exploited. This is one of the counts on which he regards the work of Santiago Calatrava as deficient. Rice tended to keep these views away from the public ear, but he did write at least one article that made his opinions clear.[56]

In his review of Calatrava's Stadelhofen in Zurich he commented on the mix of steel and concrete in the structure, saying that *'there is no clear identity, and, I believe, the inherent character of either material is not very important to Mr. Calatrava'*. He regarded it as much less proficient in concrete than in steel and found that Calatrava's work lacked the simple visual tension that made the work of

Maillart so exciting. He found Calatrava's cave-like forms at Stadelhofen lacking in refinement, and suggested that the reason for this was that '*he has not yet found the sculptural and structural contradictions of concrete*'.[57]

Lloyds was part of Rice's investigation into exploiting the sculptural and structural contradictions that he referred to. But, unlike Calatrava, he saw the clear advantages of having a dedicated architect as the conceptual force driving the design idea rather than some kind of hybrid architect–engineer. The lessons learned at Lloyds on the articulation and quality of the concrete elements would be reapplied in collaboration with other architectural partners on projects like the stadium at Bari eight years later with Renzo Piano.

Fig. 8.10. Lloyds at night.

Photo: Phil Berridge

Chapter Nine
Structural fabrics

Chapter Nine
Structural fabrics

Introduction

It was while working with fabric structures alongside Frei Otto in the aftermath of the Sydney Opera House that Peter Rice developed his interest in materials and unconventional approaches to lightweight structural solutions. It was this kind of structure which led him to be a key part of the development of the computer-based analytical technique called Dynamic Relaxation with colleagues like Alistair Day. The program was to become an important tool in his later projects. So, in a way, it might appear surprising that he did not work with fabrics more than he did.

But there is an explanation for that. He could see only limited application for this kind of material. He was only interested in using fabric if he could see that certain unique qualities could respond to or enhance the architectural intention. There were two particular qualities that he liked about fabrics; firstly the translucency. In cases where they could provide shelter from the sun or rain but usefully transmit daylight they were particularly good. Bari Stadium (chapter seven) is a case where this kind of quality could be taken advantage of to remarkably good effect.

The second property of fabric that appealed to Rice was its ability to define unique forms and spaces. The cutting patterns and resulting seams in the fabric are

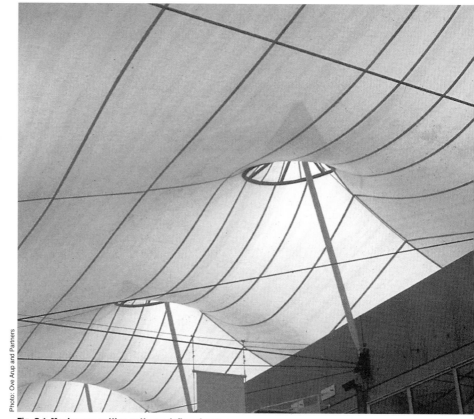

Photo: Ove Arup and Partners

Fig. 9.1. Montrouge: cutting patterns define shape and internal quality.

Fig. 9.2. Montrouge: beneath the fabric canopy.

Photo: Ove Arup and Partners

nearly always very visible in the completed structure. By careful management of the cutting pattern and the way that the edges (seams) defined by this cutting pattern in turn define flowing, doubly-curved forms, these fabric structures, and in particular the spaces below the fabric, acquire particular and intriguing qualities.

An example of this kind of quality can be found in the tent designed to cover the garden entrance area for the Schlumberger corporation in 1980: Schlumberger Montrouge (Fig. 9.1). This was to be an entrance canopy, so as a result was relatively long and thin, 100 m x 15 m in plan. The intention was that it should not only provide a cover but that it should also enhance the idea of pedestrian flow and procession. To do this the cutting pattern is manipulated so that the fabric runs in long continuous strips along the length of the canopy. Rice found the resulting form and the way that it transmitted light very pleasing.

In fabric structures the integrity and resistance to deflection is nearly always attained through shaping the fabric into an anticlastic form; one which, like a saddle, curves in two opposite directions at right angles. To do this there must be a compression element pushing the surface upwards. This compression element can be one of two major types. It can be a line support as in the ribs at Bari Stadium. Or it

can be provided as a point support, by masts that push against the ground or by 'flying struts' that are part of a net of tension and compression elements. The Schlumberger Montrouge tent uses a combination of mast and flying strut. The two further examples of Rice's fabric structures in this chapter are Lords Mound Stand in London and Les Nuages at La Défense in Paris. The principal compressive elements in these two buildings are the mast and the flying strut respectively.

Lords Mound Stand

In 1985 the architect Michael Hopkins was commissioned to design a new stand for part of the famous Lords cricket ground. The conventional way to provide cover for a stand at a sporting venue at that time was to have some kind of structural steel arrangement supporting profiled steel sheets. Those breaking with convention might choose long precast concrete beam structures. But Hopkins had a very different vision of what was appropriate; one that brought much subsequent and deserved praise.[58]

In most of the fabric structures that Rice worked on the material that he employed was PTFE coated glass fibre; what Rice often referred to by the abbreviated and somewhat deceiving name of Teflon. The coating protects the structural material from strength loss due to UV degradation in

sunlight. For the Lords Stand project cost restrictions meant that PVC was used instead.

Hopkins chose a series of white masted tents rather than their more conventional counterparts. The tents are light and flowing structures that have become synonymous with Lords and with cricket more generally. They not only do their job; they also conjure up all the right kinds of image.

The tents take up a conical form by having one single main point of compression to give the high point in the curved fabric. The high points are provided alternately by a circular hollow section mast and a ring suspended by cables from the adjacent masts. This suspension system means that a

Fig. 9.3. Mound Stand project nearing completion.

Photo: Ove Arup and Partners

Chapter Nine
Structural fabrics

mast is not required for each conical tent and thus the area beneath alternate tents is column free. The ring is required at the head of each cone to pick up the fabric in a way that does not cause heavy stress concentrations. If the fabric were picked up at a single, small isolated point at the head of the mast there would be a severe concentration of stress here, like the point of a pencil being pushed through a cotton handkerchief.

The rings in the head of the conical form help to define a space below, but what Hopkins and Rice wanted was for these sub-spaces to be as well defined internally as possible. The cutting pattern is therefore arranged to radiate from the centre with the panels of fabric getting wider at the edge of each conical form. In order to pull the lower edge of the material out into the desired shape there are main spreaders, also of circular hollow section, which are pinned to the masts. These run out at right angles to the mast and attach to a series of cables that are threaded through the pocket formed at the lower edge of the fabric. To the viewer in the stand these are seen as a free edge, made up of a series of shallow cuspate forms.

From other parts of the ground the tents at Lords read as a united whole but to

Fig. 9.4. Rice's favourite view: the fabric defining a particular internal quality.

the occupant of this part of the stand the spaces formed and the elements of the composition are read at a personal scale. Like the stand at Bari, Italy, a year later, the spectators are treated as a group of individuals, not a crowd. In subdivision and in detail, the human scale remains of prime importance.

Les Nuages, Tête Défense, Paris

The Grand Arche at La Défense in Paris is a massive, monolithic piece of architecture. It is an immense shining cube with a rectangular hole punched out of it to form a gigantic arch that marks one end of the 'historic axis of Paris'. It is the most recent of three triumphal arches on the Louvre-Champs Elysées axis, but its proportions are colossal compared to its older counterparts at the Carrousel and Etoile in the city. The Louvre, many miles away in the heart of Paris, is bisected by the axis, but is still visible from La Défense.[59] The arch was a competition winning design, for which Kisho Kurokawa was one of the judges. It is an almost perfect cube with 110 m long sides and a hole punched though it that is 93 m high and 70 m wide: tall enough to easily accommodate Notre Dame cathedral, and wide enough to accommodate the Champs Elysées.

Designed by J. O. Spreckelsen, the concept for the arch includes the idea of 'clouds' (Nuages) that would be seen as drifting through the punched hole in the arch, and spilling down the long, linear esplanade that directed the viewer to the Arc de Triomphe and Paris. To Spreckelsen these clouds were to be realised in the form of large, but apparently light, fabric structures.

In the end cost restrictions meant that the Nuages had to be concentrated around the Arche. Those conceived for the esplanade, or Parvis to give the area its correct name, have never been constructed although they were designed by Rice and his team. The esplanade as a consequence feels rather bare and expansive without them.

But the Arche itself and the floating clouds around it remain as a powerful and impressive sculptural statement. The two elements of the composition play a complementary role. The Nuages are formed in overlapping layers that Spreckelsen wanted to be read as a fluid form that would contrast with the pure rectilinear geometry of the monolithic Arche.

Unlike other projects where Rice might have to probe and extract ideas in a design meeting to establish the concept that drove the architect's ambition, here the task was much more straightforward. This was a project about sculptural ideas. It was about having a clear and explicit concept that was

Photo: André Brown

Fig. 9.5. La Grande Arche framing Les Nuages with Parvis leading to the arch.

Fig. 9.6. The fabric, tension net and flying struts working as a complementary system.

Photo: André Brown

Chapter Nine
Structural fabrics

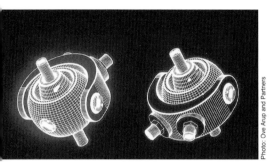

Fig. 9.7. Computer model of special joint details.

Fig. 9.8. Edges of the fabric structures and the sub-panels or 'soft volumes'.

Fig. 9.9. La Grande Arche.

to be made manifest in the built form. The fabric structures that Rice was to be asked to design were to be fluid, cloud-like forms drifting through the arch. The architectural intention was clear and the engineer could get to work.

In fact Spreckelsen became ill during the Défense project, and eventually had to resign. The French architect Paul Andreu took his place, but Rice felt that he understood Spreckelsen's intentions clearly enough to proceed with confidence.

The first line of thinking that Rice adopted was that although the Nuages were

conceived as flowing and free form in shape some standardisation was needed. In addition the soft volumes that real clouds occupy needed to be reflected in some way. The third and final major factor that needed attention was scale. The size of the opening in the arch was truly monumental and, in Rice's mind at least, a device was needed to give these large fabric structures a more human scale.

To address the first of these matters Rice adopted a modular approach. The main supporting cables radiate out from a curved inner main cable (Fig. 9.8). This produces a series of approximately parallel cables, which are then subdivided into modules. Each of these modules is based on a common 'character' so that what is produced is a limited range of variants. The limitation is dictated by the desire to have a fixed set of component parts but with enough variation to respect the style and rhythm of the clouds. Through an iterative process, and generated by a geometric rule, Rice developed a module that could be connected to itself in a variety of ways. And to reflect the fluctuations in the cloud surface the module shape is generated from sinusoidal curves, but with the shape being generated in a skewed direction.

Rice likened the process of producing the shapes of the modules to that of constructing fractal images: it is also similar to the techniques of parametric variation in the shape of forms that were used to generate structures like Waterloo Station in London. For the Nuages the principle used was that, based on the last shape formed, a new shape is generated which connects to the edge of the first shape and which derives its own shape from a characteristic set of rules.

The computational techniques that were used to derive the forms in this way were second nature to Rice. Manipulating the forms to optimise the shapes of the component parts, and the stresses in the fabric and cables was something that Rice would do himself whenever he had the chance. Seeing the results of the computations was another design device to be played around with. His partner at RFR, Bernard Vaudeville, recalled Rice scrutinising an early computer analysis for the Nuages. To most of the people in the office the result was a blizzard of numbers; but Rice was able to see the regular crystal form of the snowflake in that blizzard. He could see almost immediately how the forms and stresses could be manipulated to produce a better conditioned result, where stress concentrations would be reduced and the potential for standardisation could be optimised.

The second issue requiring attention was that of reflecting and representing the light fluffy volumes that real clouds take up. One possible line would have been to make the modules as two layers of fabric, but there was an immediate problem with this approach. The fabric chosen was Teflon, which is not particularly translucent. If two layers had been used then the degree of translucency would have been almost negligible.

The tension in the fabric of the modules was to be maintained by spreaders; thin struts in compression pushing between an inner and outer layer. If both layers could not be in fabric the logic was that one layer could be in fabric whilst the other one was made up of a thin net of tension cables in the form of irregular inverted pyramids.

Doing it this way the intention was that the translucency could be maintained, and that the transient, ephemeral nature of and the soft bulk of natural clouds would be suggested by the space between the net of cables and the fabric. Critics were divided on the success of this approach. Many felt that the fabric structures were rather harsh and solid given the design brief, but Rice felt that with age the fabric mellowed and this gave a more satisfying look and feel to the structures.

The last of the key ingredients mentioned above was that of scale, and to address that Rice turned to his old ally, the detail. Each of the major intersection and anchor points is designed with care. Typical intersection points avoid using standard flat plate solutions. Instead joints are made so that they soften the junction; here forms like an oblate spheroid and annular rings give the crafted and individual feel that Rice aimed for (Fig. 9.7).

On reflection, the criticisms of the fabric clouds, Les Nuages, are partly justified. The final form could be said to be rather flat and tense. Close to, the dominant feelings are to do with tension and engineering rather than lightness and ephemerality. There is a possible reason for this: the departure of Spreckelsen from the project due to his deteriorating health.

Rice's work was always best when it arose from a continuing refinement in response to the architect's criticism. He thought that projects were most successful when they matured as each of the collaborators contributed in turn to reaching the optimum goal. One of his pet sayings was that 'Genius is great patience'.

In this case the visionary architect whose ideas Rice was aiming to turn into reality was not there to feed from and interact with. The result is a design that does not have the architect's vision to reject engineering proposals. This, after all, was the basis for Rice's scathing criticism of Calatrava's work. The architecture and the engineering should be there as distinct ingredients in the mixture, not as a bland composite.

Chapter Nine
Structural fabrics

Photo: André Brown

Spreckelsen might well have appreciated that it was a very sophisticated engineering solution that was being proposed, but would have seen too, that the forms were perhaps more severe than he would have wished for, and the lightness was not being achieved. The replacement architect for Spreckelsen on the project, did not have Spreckelsen's insight into how the two powerful forms, La Grande Arche and Les Nuages, should be seen to interact visually. What Rice was left with was drawings but too few words. For Peter Rice the words defined the project. The architectural critique inspired the engineering. Without the informed and passionate architectural driving force the design team was incomplete.

Ian Ritchie says that Rice, as project leader for Les Nuages, asked for his critical comment on the design proposals. Ritchie's reaction was that he felt it was difficult to achieve the qualities of floating and lightness within the constraining window of the arch. But Rice remained optimistic. He felt that with age and maturity the Nuages became more satisfying and successful. They certainly have an engaging quality in their own right.

Chapter Ten
Glass and steel

Chapter Ten
Glass and steel

Peter Rice felt very strongly that the Victorians had succeeded in producing buildings in glass and steel that had a crafted and cared-for quality that we regularly fail to achieve today. With the emphasis now on speed of construction our industries have been led by the nose down the narrow alleyway of standardisation and conformity. Rice knew well enough that the result was a design environment that was hamstrung and limiting. He sought every opportunity to break free from the shackles of standardisation, or industrialisation, as he often termed it.

He looked back at the work of the Victorian engineers and reflected that

'We have learned so much about steel and glass and how structures work since then. Where has the knowledge gone? The language of the industrial product, the Section, the Tube, has dominated the individuality. The joy and the delight have been smothered.'[60]

In his glass and steel structures Rice found ways of challenging the conventions and offering a range of new approaches that arose from his ability to return to first principles without fear, and without the need for the crutch of a code of practice. He regarded this as a time of opportunity; a time to experiment. This is the period, he felt, when we should be breaking free from the grip of industrial monotony and mechanical repetition.

Clearly buildings like Beaubourg, and more importantly in this respect, the Bioclimatic Façade at La Villette, were the testing and proving grounds for Rice's ideas in glass and steel. But if the search for the source of the stream started there an important influence would be missed. That influence was Sydney Opera House. When people talk about the Opera House it is nearly always the ribbed concrete structure, the clever 'sections of spheres' solution and the tile lids that are the focus of attention. To most it is a concrete building. This means that other aspects of Utzon's influences as a talented designer and the Arup team's ability to make his ideas work are missed. One major element that falls into this arena is the glass walls.

The glass walls for the main auditorium shell at Sydney face towards the city and offer a wonderful view of the commercial area of the city around the harbour. As site engineer, Rice was just as responsible for these elements as he was for the concrete main structure. What was needed was a very large glazed end wall for each of the shells. The solution that was devised was inventive and interesting and must, surely, have been provoking thoughts in the mind of Rice. John Nutt remembers Rice as being 'meticulous in detail at the Sydney site'.[61]

The main supporting structures for the glazed walls are specially fabricated steel beams that generally span vertically from the ground to the edge rib of the shell. The beams are made by welding two small hollow tube sections to a steel plate. In effect the tubes are the flanges, and the plate is the web of the beams that are formed. Steel connectors are fixed to one of the tubular flanges and these connectors, which are adjustable, pick up the glazing.

Many years later Rice was to develop systems that were much lighter in appearance than the support structure at Sydney. But Sydney had several of the important ingredients of a Rice glass and steel wall structure that would be developed later; the structure set back from the glazed plane so that the glass reads as a transparent sheet, the machined and crafted connector that extends from the beam to the glazed plane, and the articulation of all of the elements in the composition. All of these features are evident in later Rice buildings. Utzon was certainly the influential figure that Rice recognised as a key mentor in his architectural appreciation.

Pyramids at the Louvre
In the later stages of the Grande Pyramide project at the Louvre, RFR were called in to assist with the construction and fabrication. Peter Rice and his teams at Arups and RFR had, by the early 1990s, developed a

Photo: André Brown

Fig. 10.1. The Grande Pyramide at the Louvre, Paris.

Photo: André Brown

Fig. 10.3. A tie and strut connection detail.

reputation for working with innovative and minimal glass and steel structures. The Grande Pyramide project would mark the public entrance from the large open courtyard down into the subterranean Louvre galleries (Fig. 10.1). These were to be structures that would make a striking contrast to the heavily decorated and ornamental stonework in the Louvre building which define the courtyard. I. M. Pei agreed that Rice should be called in to make sure that the technology would make the powerful architectural idea work. The pyramid was to be about sharpness and defined by pure geometry.

Rice clearly liked Pei's intervention at the Louvre for the bold sculptural statement

Photo: André Brown

Fig. 10.2. Looking out from the Grande Pyramide to the classical Louvre buildings.

Chapter Ten
Glass and steel

that the pyramids made. Later, in 1990, when Rice and Pei worked in collaboration on the inverted pyramids there seems to have been considerable admiration on the part of Peter Rice for the design talent of I. M. Pei. The tension created by an inverted glass pyramid balanced, point to point, on the marble pyramid below, at a crossroad of circulation routes through the gallery, appealed immensely to Rice. He could appreciate the architectural intention and the architectural quality.

Consequently, although he might have wished inside that Pei would dispense with a potentially redundant sub-frame of aluminium glazing supports as in Rice's later project at Charles de Gaulle airport, he was willing to work with Pei to produce what he believed to be a fine architectural statement. He was appreciative, not dogmatic. As with his work with Tschumi he found excitement and potential in the bold sculptural solution. For Rice the tension and instability that he looked for could be provided by visual means as validly as it could be from structural means.

In many of Rice's schemes the idea of layering, as in the main trusses and sub-trusses at Les Serres in La Villette, was used as an articulating device. But Rice noted Pei's 'horror' at such techniques and was prepared to accept this view in order to make Pei's strong visual message work in the way that the architect intended.

In this respect it is worth commenting here on Pei's design of the Grande Pyramide at the Louvre. Looking at the kind of solution preferred by Rice it is likely that he would have suggested removing some of the cable elements from the inside face of the tensile space structure. For the Luxembourg Museum designed in 1990, for instance, Rice took the idea of a set of starburst cables as restraining elements which meant that the inside face no longer had to do the job of helping to resist wind uplift. A variant of this idea could have been adapted for the pyramids at the Louvre. But it was out of respect for Pei's architectural skills that Rice would have been able to forego his strong preference for lightening the grid of tension members on the lower face, possibly in conjunction with a layered structure.

I. M. Pei, working with Michel Macary in 1985, also collaborated with Rice on the glass and steel structures inserted at roof level over the courtyards in the Louvre. In both the Grande Pyramide and the Louvre courtyard schemes Rice had demonstrated an ability to realise glass and steel structures of great quality. He had also shown his ability to interpret the architectural idea and to deal with the practical issues in a way that would support the idea. So when the Pyramide Inversée project came along (Fig.10.4), naturally Peter Rice was the preferred collaborator. The Pyramide

Inversée project does show the kind of articulation and layering that Rice was so taken by, and the evidence is that Pei had come to appreciate that Rice's approach could enhance the architectural quality of this kind of scheme. There is certainly a difference in the technological approach taken for the Pyramide Inversée compared with that of the Grande Pyramide.

In terms of the engineer's contribution to architecture the nature of the evolution of Pei's work clearly reflects the significant influence of Peter Rice and his ideas on how technology and architecture can evolve together. If more evidence of this is needed we only need to look to the fact that I. M. Pei's son D. D. Pei went to work at RFR where he became an associate. There he worked with Peter Rice and Henry Bardsley on the glass and steel structures for the Museum in Luxembourg, a project that was to be one of the last that Peter Rice would work on.

For the Pyramide Inversée scheme, as with all his other projects, Rice saw his role as being to facilitate the architectural idea, not to oppose it; to make the idea work the way it was conceived rather than to challenge the conception. Structure and technology would be enrichments and enablers, not as prescriptors or constraints. The fact that Pei's approach was different to that of Rogers or Piano simply meant that

Photo: André Brown

Fig. 10.4. The Pyramide Inversée (inverted pyramid).

Fig. 10.6a. Apparent complexity in the tension cable and flying strut system.

Fig. 10.5. Computer model of the pyramid.

Photo: Ove Arup and Partners

Photo: Ove Arup and Partners

Fig. 10.6b. Connections and subdivisions.

Photo: André Brown

the way the design could be enriched and enabled would require Rice to take a somewhat different role. But he did this with his eyes open. He was candid that Pei was a different kind of architect, and the architectural product was the evidence that the different approach was valid.

The inverted pyramid was the device to be used to cast and reflect daylight into the underground carousel gallery of the Louvre. This is a gallery space containing exhibition areas and shops, and the pyramid is located at a kind of crossroads in the gallery. Being a square based pyramid means that there is an angled face of the pyramid pointing in the direction of each of four axes. And because there is more light coming from the sky above than from the gallery below, the glass sheets in the pyramid act as good reflectors, bouncing skylight into the gallery spaces. This gives the pyramid a very particular quality. Rather than being simply transparent for much of the time it actually appears luminous. For this reason, it is also referred to as a 'lustre' (a chandelier).

But Pei's intention here was not simply driven by the need to redirect the daylight. Like the Grande Pyramide the rectangular hole which let the light into the gallery was seen as the framing device for a strong sculptural object or, more correctly, a family of objects. The pragmatics of the situation meant that these objects were, like their predecessor, to take the form of a pyramid. These pyramids were aligned along a common vertical axis.

The upper skin of glass is at ground floor level and is a very flat pyramid, the shallow slope serving to shed rainwater. (Fig. 10.7) Although this skin is not intended to have pedestrian traffic passing across it, it had to be designed as a floor that could be fully loaded. This is because only a low wall at ground level protects it. Like the rest of the pyramid, the structural support is afforded by a tension structure, with discrete struts, that becomes deeper at the centre. The structure defined is therefore in the shape of a convex lens.

The structure supporting the lower faces of the pyramid, the inverted pyramid from which the project gets its name, is a tensegrity structure, i.e. any compression struts are discrete and located in position only by tension elements. The structure therefore gains its integrity through a continuous web of tension elements: hence its name, as coined originally by Buckminster-Fuller. Rice referred to the structure as having main support cables and 'flying struts'.

In effect, the triangular faces of the pyramid are structured like the walls at La Villette, but the tension is provided by the self-weight of the glass walls that hang from the concrete ring beam which defines the square hole at ground level. Because the inverted pyramid is sheltered from environmental loads additional pretensioning beyond that provided by self-weight is not required.

These curtains of suspended glass are then drawn together to form a pyramid shape by thin cables which attach to the main cable–flying strut system (Fig. 10.5). The whole system reads as a series of 'volumes' and the glass face of each volume is attached to the volume immediately above by means of a crucifix shaped connector (Figs. 10.4 and 10.6b). This connector both transfers the dead load that provides the tension, and it prevents rotation at the glass interface of the volumes. To accommodate this function, at each of the four ends of the crucifix there is a connector that attaches it to the appropriate glass diamond shaped panel. As with other aspects of the structure this connector is part of an evolutionary process; it is based on, but moves on from, the special spherical bolting system developed for La Villette.

Fig. 10.7. Louvre pyramids: top face of the Pyramide Inversée in the foreground and Grande Pyramide in the distance.

Photo: André Brown

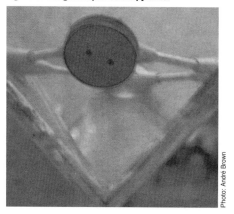

Fig. 10.8. The glass apex of the pyramid.

Photo: André Brown

Photo: André Brown

Fig. 10.9. Grand Ecran (Big Screen) building.

Photo: André Brown

Fig. 10.10. Junction of the glass and steel wall and roof structures.

Grand Ecran

In the Grand Ecran (Big Screen) project Rice and his engineering colleagues at Ove Arup and RFR worked with the Kenzo Tange architectural practice and with artist Thierry Vide. This is one of a suite of projects that followed from the reputation gained at La Villette in which the potential for extension and variation of the idea of marrying a thin steel tension field system with a plane of glazing was developed.

Of course the opportunity to work with a practice with the reputation of Tange and the involvement of an artist in the sculptural aspects of the work on the entertainment centre was an additional bonus. But projects like this allowed the

ideas and principles set in 1981 at La Villette to be extended and modified.[62]

In the offices for the Bull Corporation in Paris, six years after La Villette, deep triangular cross-section trusses had been set side by side to form an atrium roof structure. And in 1991 restrained bow shaped transom trusses were adopted for the glazed wall structure that was part of the atrium glazing system developed for 30 Avenue Montaigne, also in Paris.

The Grand Ecran in Place d'Italie, Paris, just predates the Avenue Montaigne project and can be seen to be a member of the same family, in the same way that the roof structure at the Bull offices can be seen to have a similar genetic make up. In the main space at

Photo: André Brown

Fig. 10.11. Tensile roof truss system.

Photo: André Brown

Fig. 10.12. TGV/RER Station: supporting fingers in the area of the railway platforms.

Fig. 10.13. Supporting fingers at airport concourse level.

Photo: André Brown

the Grand Ecran, the glazed wall structure is supported by some impressive tensile gymnastics (Fig. 10.10) reminiscent of La Villette. The roof structure uses a triangular cross-section that helps to resist lateral instability or torsion (Fig. 10.11). The glazed surface on the top plane of the triangle is a plane of stiffness that contributes to the torsional rigidity. In this way, it can be thought of as an upside down version of the Japan Bridge (chapter eleven) which also relies on a triangular cross-section with a stiffened plane to counteract potential torsion.

Humans in transit

TGV/RER railway station roof: Roissy Charles de Gaulle, Paris

Railway stations had been a test bed and source of innovation in the development of long span structures in the early days of iron and steel. The naval architect Richard Turner at Liverpool Lime Street and Barlow and Ordish at London St Pancras had demonstrated in a grand style how great cathedrals of glass and steel could be created with techniques that had not been possible in other materials. They took the potential made evident at buildings like Crystal Palace and extended the ideas by applying new forms and new ideas in the context of the needs of a major railway interchange. These were the

kind of pioneering designers that Rice drew inspiration from. They invented, they thought laterally and they mixed experience and expertise with their inventiveness to evolve ideas that were, at the same time, both innovative and pragmatic.

Peter Rice developed a reputation in France for devising glass and steel structures that were daring and imposing. At Roissy Charles de Gaulle airport the high-speed railway line traverses the airport by being set down below both the runway and road levels. Where the concrete tunnel structure ends at the station, RFR were asked to design a glazed roof and walls to cover the platforms in the Rice tradition. Aéroports de Paris and the French railway company SNCF were the official architects for the scheme but, in reality, RFR were the designers in both the architectural and engineering sense.

On the face of it this would have set up an unsatisfactory working condition for Peter Rice, where multi-disciplinarity contributes, essentially, to the richness and success of a project. The answer is that the practice of RFR is a mixed environment that brings together both architects and engineers. Consequently, even though it was one professional organisation, the architects and the engineers were able to exchange ideas and evolve the design in a framework that gave both professions their say. In fact this particular way of working had the particular

Fig. 10.14. Cast connector at the base of the raking columns.

advantage that the architects and the engineers shared a common philosophy based on the benefits of close working with their collaborators, from the start of and throughout the whole project. That is why they were at RFR, after all.

The architectural concept had two particular strands that would guide the direction of the project. The first was that the station should contrast sharply with the dark enclosed tunnel space that emerged into it. The experience of moving from darkness to natural light should be emphasised. This meant that the structure had to be very light and transparent. Looking up to the sky should be as important as any structure that might be up there in the roof.

What the RFR architects also wanted was to make the roof appear to float over the cutting that the station sat in, making reference to the machinery of flight to which the airport owed its existence. The separation of the station structure from that of the cutting and tunnel was a way of defining a distinct character for the station, so the support for the roof should not come from an extension of the cutting walls. Although the enclosure would require the walls above the cutting to be glazed those walls should gently touch, rather than support the roof that they met.

The idea of articulation that was close to the Rice heart was to be applied here in a way that revealed an organisational principal that was stronger than in most buildings. With Piano the idea of articulation was relatively flexible and dealt with project by project; the intention simply was that the pieces from which the building was composed should be read as distinguishable elements with their own character. In the TGV station the articulation is taken a stage further and is used as a system to organise the structure and construction. The layers of the structure are superimposed in a deliberate fashion, so that they can be read as a set of distinct 'virtual surfaces', the elements in those surfaces becoming lighter and more slender with distance from the ground.

This was not a small project; 400 m long in total at a cost of 120 million francs. It

Fig. 10.15. The floating roof structure.

Chapter Ten
Glass and steel

Photo: André Brown

Fig. 10.16. Double lower boom in the bowstring truss.

would be a major and imposing form. Between it and Terminal 3/2F the two projects would redefine the quality of this major transport interchange for the French capital. The oppressive dullness of Terminals 1 and 2 would still be there but their presence merely emphasises the lightness and quality of the Rice buildings.

The platform level is some distance below the concourse level. At the concourse level the supporting structure takes off from a low concrete plinth. Over the platform the equivalent of the plinth is extruded downwards to meet the concrete foundations below track level. In cross-section the structure is supported by a pair of masts that rise from the plinths at approximately third points in the span. So, roughly speaking, the structure has a centre span that extends outwards at each end to form long cantilevers that are at least one third of the total length of the roof truss. Consequently, the dominant tendency is for the roof structure to hog (bow upwards) over the supporting structure, putting the top face of the truss into tension: the reverse of what is normally the case. This means that the top face can be made as a relatively thin tie, while the lower boom needs to be made with sections of a larger cross-sectional area to resist compression. As in all cantilevered structures the bending forces (moments) are greatest at the support point and taper off to the ends of the roof structure.

The form of the roof truss developed to respond to this structural arrangement is a very deep crescent, like a giant longbow, slung from the top of the supporting masts. These crescent-beam trusses, or 'poutres-croissant' to use the French terminology that Rice preferred, are tied down at their ends to prevent wind uplift which would cause a reversal of the loading conditions described above. This means that the condition of compressive lower boom and tension upper boom is maintained in all loading conditions, and the structure can be lighter along the top edge, in line with the guiding principal of

Fig. 10.17. Separation of the wall and roof structures.

Photo: André Brown

Fig. 10.18. Spreaders pick up the corners of the glazing panels.

Fig. 10.19. Diffuse quality of light beneath the structure.

hierarchical layering. In fact the lower boom is made as a pair of hollow sections, joined by welded stubs, to emphasise the size in relation to the top (tension) boom, and to give better lateral stability.

The masts that support the poutres-croissant are circular hollow sections, which read as fingers with the thin top chord of the crescent beams slung over their ends. They are arranged in groups of two, three or four, rising from a single plinth, and radiating out so that each of the masts supports a different truss. At the plinth the focus of the radiating masts is dealt with as a cast steel piece, a semi-circular element on a flat base that carries on the analogy of the slender supporting fingers, uniting the fingers together like a part of the hand. The cast nature of the piece means that this unit softens the rigid geometry of the circular hollow mast at this important visual junction. And being the element that the passenger

brushes past on the journey through the station it was important to Rice that the trace of individuality and craft should be particularly evident in this element.

One criticism of the column base junction is that the masts are actually pinned at this support, whereas the junction actually reads as a rigid joint with the mast section flowing into the supporting piece. To accomplish this the thinner section at the pin is hidden by a semi-rigid compressible filler. It is untypical of Rice's approach; to fool around with a junction in this way. His usual approach would have been to find a way of making the technical function explicit while maintaining the architectural (in this case visual) concept.

Fig. 10.20. The railway station-airport interface.

123

Chapter Ten
Glass and steel

The glazing on the top face of the building is supported, some 3 m above the top edge of the crescent trusses, by slender struts that extend up from alternate nodes in the top boom of the trusses. By decoupling the structure and glazing in this way the glazing could be applied without the need for construction joints. Any movement in the main structure could be accommodated by rotation in the ends of the supporting struts. The second advantage of this arrangement is that an unobstructed space is created between the truss and glazing that allows access for a cleaning gantry.

Fig. 10.22. Looking down the axis of the building towards the boarding gates.

Fig. 10.21. Roissy Charles de Gaulle Airport.

Photo: André Brown

Photo: André Brown

Typically, light was also a particular aspect that was handled deliberately and successfully in the station structure. Fritted glass panels are used to soften the quality of the natural light during the day. At night the artificial lighting is bounced off the fritted glazing to bathe the platform in a diffuse reflected light. The quality of the light, and the way it interplays with the white structural elements is a real success; it gives the building a cool refinement and great quality.

Terminal 3/2F: Roissy, Paris, Charles de Gaulle

Terminal 1 at Roissy Charles de Gaulle, built in 1974, is, like its later extension, Terminal 2, a heavy, and oppressive affair of little architectural quality and grace. Martin Spring was very astute in his description of the pair of them: 'an elephantine celebration of France's favourite structural material, in situ reinforced concrete.'[63]

The 1990s addition that Peter Rice and his co-workers were responsible for is a completely different animal; a light and delicate demonstration of how technology, function and aesthetic appeal can sit comfortably side by side. At its inception the project was called Terminal 3, but is now known as 2F. It is the terminal that is used by the host country airline, Air France, for obvious reasons. It exudes efficiency and style.

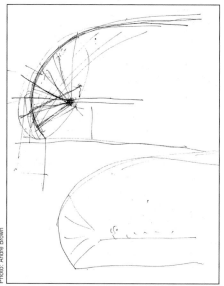

Photo: André Brown

Fig. 10.23. Rice's sketch: ensuring visibility of the radiating arms that support the skin.

Photo: André Brown

Fig. 10.24. Openings in the concrete shell.

Fig. 10.25. Supporting arm for the glass seen through the perforated concrete shell.

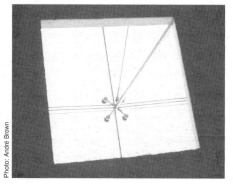

Photo: André Brown

Fig. 10.26. Articulating arms that support the ribbed structure.

Photo: André Brown

Chapter Ten
Glass and steel

The main arrival hall is a thin, pristine, curved concrete shell structure punctured by a carefully composed set of rectangular holes that give glimpses out to the runway. This curved structure is sliced in two by the embarkation–disembarkation peninsula; a 200 m long light, curved, translucent form that stretches out to the runway to receive up to 14 docking aircraft.

A long spine runs through the arrival structure from the arrival hall to the end of the peninsula. This spine beam acts like the keel of an upturned boat and the remainder of the peninsula structure continues this boat analogy in both shape and structure. It is formed from a series of symmetrical transverse steel ribs, spanning 50 m at the widest section of the peninsula, but getting shorter towards the furthest boarding gate;

Fig. 10.28. Looking down the building axis.

the prow. The structure is continuous and has no expansion joints along its entire length. The main spine helps to restrain movement but Rice was intrigued by the possibility of allowing articulation in the structure to accommodate additional movements. His sketch (Fig. 10.23) shows that he was interested in both the action of the technology by allowing movement through articulation, and the visual impact of the arm connections to the main structure (Fig. 10.25 and 10.26b).

The structure is minimal and light. The transverse steel ribs are supported by the fan of tension rods at the floor level. These ribs are a solid curved section, as they spring

from their supports at floor level. Just above head height in the main concourse, they turn from solid into trussed elements, their depth increasing as they rise to meet the longitudinal spine beam. It grows like some kind of delicate organic form between its supports at the base and the steel box section spine. The whole structure forms a doubly-curved, finely woven web of steelwork, which, despite its complexity, retains a visual clarity that

Fig. 10.27. Steel skeleton and concrete shell.

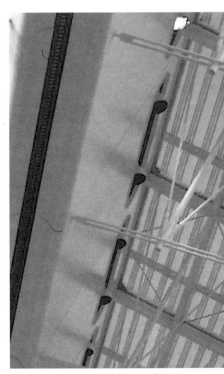

Fig. 10.29. Terminal 2F Charles de Gaulle. the trussed ribs meet the main spine.

is striking and dramatic. To get to the position where the structure as a whole could be optimised RFR developed a computer based non-linear analysis of the whole structure. The output could be viewed as a 3D model in which colours were used to indicate the stress levels. This gave an immediate visual picture of how the stress patterns had responded to changes in the structure and where the critical points were located.

The Terminal 2F structure is a real success. It is functionally effective in the way that it handles passenger movements, but the transitions from one form to another, the filigree nature of the steel ribbed structure and the explicit detailing make the journey through the building a delight.

Kansai Airport

The Kansai Airport project presented another welcome opportunity for Peter Rice to work

Fig. 10.30. Model of Kansai: Renzo Piano.

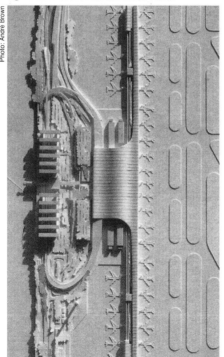

Photo: André Brown

with his long time associate, Renzo Piano. The airport project was awarded following a competition in 1988, and the project was completed in 1994. The site was unusual. An artificial rectangular island, 1·5 km x 4·37 km, had been specially created in the sea. The precise and deliberate nature of the site meant, for Piano, that the project was about producing a precise and functional machine.

In cross-section the roof structure of the main hall is an aerodynamic shape with undulating forms; the shape being created by a series of triangular cross-section trusses. Those trusses are delicately perched on a pair of raking struts at each end, like the finger and thumb of each hand holding the roof structure aloft for inspection. It is easy to envisage Renzo Piano holding the model like this to test its balance, and to test his satisfaction with the final flowing shape.

The form is derived from the concept. Piano saw Kansai Airport as a broad winged bird, or a plane coming in to land. Rice understood the image and added to it himself. He saw it, in detail at least, as a Bleriot plane; a combination of craft and engineering. The respect that he had for the craft and invention displayed by pioneering engineering designers was not limited to the world of buildings. The Bleriot plane was an object that combined elegance and engineering in a way that pleased Peter Rice.

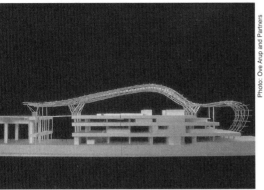

Fig. 10.31. Sectional model showing transitions landside to airside.

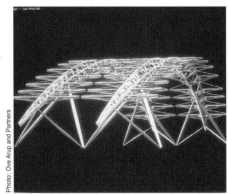

Fig. 10.32. 3D model of roof structure.

Fig. 10.35. Physical model of the sweeping roof structure.

Fig. 10.33. Airflow testing.

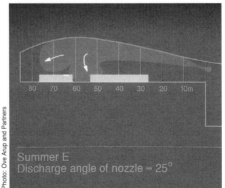

Summer E
Discharge angle of nozzle = 25°

Fig. 10.34. Computer modelling of the airflow.

It was just as valid a source of inspiration as any other designed object or artefact.

Another theme that runs through Rice and Piano buildings is the human scale, and the cross-section works in a way that supports this idea. On the land side of the building there are three external levels for pedestrian access and exit. This means that at the top storey the cantilevered 'tail fin' of the roof structure projects out over the top level at a height that is appropriate for the reception of people and motor vehicles. At the planeside there is a single level for exiting to the planes, so here the building reads as two to two and a half storeys high.

The inverted pyramid structures that support the roof come down to meet the floor on the upper level. The loads from the base of these pyramids are then transmitted downwards through concrete columns that penetrate through each of the four levels below, to be supported in the basement structure. This means that the main column and floor structure also acts as a bracing device for the roof support pyramids.

Fig. 10.36. The completed building.

The aerodynamic trusses cantilever out both at the landside and the planeside of the structure, and have a principal span of 82·8 m. The flowing form of these long trusses responds to three factors. It supports the overarching architectural concept that the building itself should be read as an aerodynamic, flowing form. The arched form also helps in resisting deflections over a long span. But the need to get good, clean air circulating through the building was important too, and Tom Barker, Rice's colleague from Arups, worked on the airflow management from the early days of the project (Figs 10.33 and 10.34).

With the body of this giant bird of a building defined by the flowing trusses and the floor levels in the volume below it, the wings then become a 1·7 km long linear extension running away from this body, left and right. Passengers enter into the body then progress along one of the wings to reach their boarding gate.

To maintain the lightweight, flying analogies a particular kind of structure is adopted for the 'wing' structure that relies on surface or shell action for its strength. Each of the wing structures dips down from the body to the wing tips so that the control tower has a clear view to the last boarding bay. The shape of the roof of the wing

structure is part of a giant toroid, which is best envisaged as a giant curtain ring 90% of which is buried vertically below ground level. The roof structure is the upper surface of the 10% that appears above ground.

The concept of lightness and efficiency, that is key to the project, is accomplished by having a very compact structure. Steel ribs define the curved profile. These connect to, and share structural action with, a set of parallel secondary beams that run the whole length of the wings and a set of diagonal ribs to complete the stiff shell form. The stiffening effect is enhanced by steel tension bars located at alternate ribs. The composite action of all of these elements in a curved shell form produces a structure that has minimal thickness to cope with the stresses, buckling effects and deflections to which it is subject. Like the skeleton of the bird, or the

Fig. 10.37. Stansted Airport.

Chapter Ten
Glass and steel

frame of the plane that it takes as its example, weight is minimised and structural efficiency is maximised.

Stansted Airport

The year around 1981 saw a rich harvest for Peter Rice. The Serres (Bioclimatic Façade) with Adrien Fainsilber at La Villette, and two of his favourite projects with Renzo Piano, the IBM Pavilion and the Menil Collection Museum all came along at the same time.

Fig. 10.38. Peter Rice's deliberations on dealing with sway and tolerances.

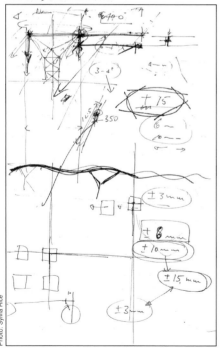

Photo: Sylvia Rice

The work at La Villette and the diversity of the other two projects led to a further complication, which was the opportunity now presented to give birth to the experimental Paris based practice RFR. Actually this was not a complication for Peter Rice, but the chance to embark on a career long dream; but it meant that time for other projects was limited.

The other project that came along in 1981 happened to be Stansted, and it is typical of the energy and enthusiasm that Peter Rice showed to note that he did make a contribution and was not simply an honorary team member. That is evident from his sketchbook. The Stansted structure was devised as a forest of steel trees; a set of almost independent vertical cantilevers of a common form that was refined to satisfy a set of very diverse criteria. As a forest of vertical cantilevers, a problem occurs with the attachment and continuity of the roof structure and cladding. Rice's sketchbook shows him working through the question of sway in the structural trees and working out how an acceptable set of deflections and tolerances could be achieved.

The structural trees at Stansted are repeated on a 36 m grid and support a roof form that is, like the trees, made in tubular steel. The roof is a single layer, doubly-curved structure that attaches to the four main branches of the tree that spread up

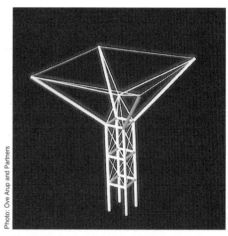

Photo: Ove Arup and Partners

Fig. 10.39. Computer model of the tree structure.

Fig. 10.40. Bracing in the tree structure.

Photo: André Brown

Photo: André Brown

Fig. 10.41. Detail at the junction of the ties.

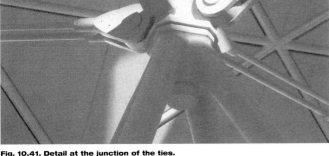

Photo: André Brown

Fig. 10.42. Roof treatment to admit daylight.

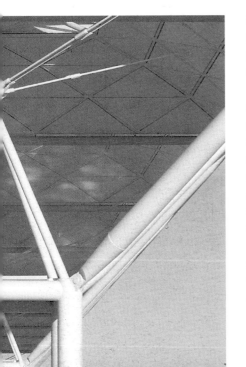

from the braced trunk. The main branches of the tree are tie braced back to the centre of the tree to keep the weight of the structure down.

In fact, one of the advantages of choosing a generic tree form for the structure meant that its design could be refined and optimised. The tube and bracing sizes and locations could be altered and the effect on the stresses, deflections and form studied (see Fig. 10.39) in more detail than if there had been greater variation in the structure across the site.

The computational analyses that were run could be tweaked to respond to the requirement to minimise the weight of the steel in the structure while achieving a

satisfactory overall form. The requirements to resist factors such as overturning, torsional buckling and excessive sway could be modified by altering the tube properties, the geometry and the level of pretensioning in the tie members.

The tree structures perform not only the primary structural function, but they also act as grid location points. As such, they house the air supply, they define the nature of the openings to admit daylight (Fig. 10.42), and they house the uplighters and the passenger information displays. In the end the project had as one of its key driving forces, and was defined by, the optimisation of the technology; hand-in-hand with the refinement of architectural form.

Chapter Eleven
Bridges and sculptures: artists and architects

Chapter Eleven
Bridges and sculptures: artists and architects

One of the clear messages from Peter Rice's work was his desire to break free from technological predictability. He found different ways and means of attacking this problem, one of which came through working with artists and architects whose approach was to experiment with sculptural forms. One consequence of this is that the resulting forms generally do not lend themselves to an ordered rationalisation of the component parts; the kind of articulation and layering of elements that is evident in Rice's work with architects like Renzo Piano and Richard Rogers. But this was not a problem for Peter Rice and, in fact, it had its advantages. If monotony and convention were the enemy it was best to wage war on many fronts. And if he was to work with collaborators who regarded their buildings as art objects then better still; this was precisely the attitude that Rice wished to adopt. It was the way that he thought about the most treasured of his design work.

Artists and designers like Bernard Tschumi and Frank Stella worked productively with Rice on structures that offered stimulation for the mind of both the artist and the engineer. But the main aim was to attract the attention of the people who saw the resulting structure. Rice wanted them to stop, look and think. For Rice it was '*less important whether they liked*

the result than that they have become involved'.[64] It was better to stir some kind of emotion than no emotion at all.

Enjoyment was the way forward in challenging the routine industrial product and in Peter Rice's case working with this group of artists certainly provided that kind of enjoyment. The environment created at RFR in Paris was all about trying to support and promote working in this way. In the office today miniature Stella sculptures sit alongside other building projects. As in Rice's mind the boundary between sculpture and architecture is a fluid one.

The four cases presented below are illustrative of the way of working and the products that came out of the collaboration between Peter Rice and particular artist–designers. In some ways it is misleading to divide the work off in this way and report it as a very different way of working with identifiably different products. Really, this kind of work simply represents the end of a spectrum, and the belief in making, and wish to make, structure and art synonymous can be found throughout the catalogue of Peter Rice's work. The examples below are simply instances that serve to illustrate the points.

Bernard Tschumi
At Parc La Villette Rice worked with the architect Adrien Fainsilber on the Serres,

the Bioclimatic Façade (see chapter six). This is a building that is about transparency and a minimal but carefully and deliberately layered and ordered structure. The minimal nature of the structure challenges the viewer initially to understand how the building manages to stand up. Closer inspection adds a further challenge; to understand why certain loadings and lateral instability do not lead to failure either.

This delicately managed approach in which the technology informs the architectural solution makes an interesting contrast with the neighbouring walkway-bridges, known as '*Galeries*', at La Villette that were designed in collaboration with the architect Bernard Tschumi. In fact Rice made the comparison himself.[65] In the Galerie structures it is the architecture that informs the structural design. Tschumi's approach was about setting up conflicts, clashes and visual contradictions, and the engineer's role was to make the overall structure and the detail sustain those contradictions.

In polar contradiction to the discipline applied in the Serres, the Galerie structures reject the use of symmetry. Supports are off to one side and rake at a variety of angles. The structure is a bit like a weightlifter straining to hold the bar (the roof) in front of the body before it is finally hoisted into the more comfortable position above the head.

Photo: André Brown

Fig. 11.1. Tschumi's walkway structure at La Villette.

Photo: André Brown

Fig. 11.1a. The Galerie structure.

One detail, which occurs at eye level, draws particular attention to the clashes that are set up in the structure. A diagonal tension element was placed by Tschumi in the same plane as the main supporting column in such a way that the tie either had to pass through or around the column. Now, making reinforced holes in the column and passing the tie through the column would have been one solution, but Rice, in conjunction with Tschumi, came up with a different one. By diverting the tie around the column by way of an almond shaped ring (Fig. 11.2) the collision between the tension and the compression elements is emphasised.

The Serres and the Galeries represent two very different approaches to the use of steel. But both had a strong architectural driving force behind them, and that was the key as far as Rice was concerned. For him *'each architect's attitude and opinions becomes part of the problem being solved'*,[66] so that preserving the conflict and visual disturbance that Tschumi wanted to accomplish became a valid basis for working.

In the end the Serres and the Galeries are both about being challenging: they each challenge convention and they each challenge notions of stability. They do this in different ways but they have this important ingredient that means these structures help to push at the boundaries of what is accepted. So Rice was happy to work with these architects,

Photo: André Brown

Fig. 11.2. Detail at a typical column-tie intersection.

confident that working with a collaborator who operated in this way would allow elements of engineering exploration to take place, since exploration and invention was at the heart of each of the ideas. With projects like the Serres this exploration would come early on and provide the team with a kit of technological parts and solutions to integrate into their design. With Tschumi and projects like the Galeries the strong sculptural idea would come first and the engineers' task would be to devise a technological solution that not only made the building stand up, but also reinforced those architectural ideas.

Frank Stella

Frank Stella was unequivocal in his praise for Peter Rice. The first time that they met on what proved to be an abortive bridge project over the Seine at Porle in Paris, Stella realised that he was dealing with an engineer who had an extraordinary talent that could take small models of conceptual forms and turn them into full-scale objects. Objects that really worked.

At their first meeting Stella, a New York based artist, showed Rice his ten foot long model of the proposed bridge: a spray painted jumble of aluminium. Before the model could be presented to the French Minister for Public Works it had to have an engineer's seal of approval; a confirmation

that it could really be built and stand up. Rice surprised Stella first by taking the project seriously and turning up to view the model as promised. Then he followed that by confirming that it was buildable.

It is revealing to note Stella's reaction to this:

'Somehow, even though he communicated a questioning, perhaps conditional, sense of approval he did it in such a way that the recollection makes me happy to this moment…just to be around him makes you want to think, think as hard as you can.'[67]

Stella realised that in talking to Rice all he needed to do was to make his idea explicit and Rice would be able to make it work. From then on Rice was to provide the technical know-how that could turn Stella's card models of curved interlacing forms into full size artefacts.

In 1990 Stella was invited by the architect, Alessandro Mendini, to collaborate on the design for a new museum in Groningen in the Netherlands. Stella was given the upper part of a wing of the gallery to design in his flowing sculptural style and produced an initial model made from aluminium pieces. When Rice quizzed Stella on the generator and concept Stella revealed that he had got the idea of the form from a book on Chinese lattice designs. Taking a leaf from these designs and twisting it gave him the form

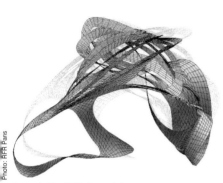

Photo: RFR Paris

Fig. 11.3. Frank Stella sculpture.

a subset of disciplined geometric forms without losing the quality of the initial concept. One of the features which is evidence of the rationalisation is that the geometry of one of the leaves is generated from the other by translating and rotating the original form to reduce variation in the elements that the form would be finally constructed from.

The second aspect of the way that Rice evolved a solution here, just as the adoption of computational techniques, is typical of Rice's general approach. He brought together a set of materials that responded to the needs of the project rather than to any preconceived notion that one material would be particularly appropriate for a project. A continuous glass wall defines the boundary of the gallery and the leaf structures of the roof overhang these walls. The 'veins', in the now rationalised leaves, are picked out as laminated timber beams and the position of the veins also defines the position of partitions below, where they are needed. Between the veins of laminated timber sits a sub-structure in the form of a grid of timber joists. Over this grid a double layer of Teflon coated fabric is stretched. As in other projects Rice works with a wide-ranging palette of materials and in the Groningen Museum project, as elsewhere, marries a set of materials from that palette in a way that enhances the architectural idea.

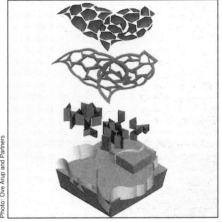

Photo: Ove Arup and Partners

Fig. 11.4. New Groningen Museum: Frank Stella's model.

that he was looking for, and the idea developed into two of these leaves that intersected.

With the core idea made explicit Rice could adopt a working strategy to make sure that the idea could become a reality. To do this he took three steps. The first was to use 3D computer graphics to model the form so that from the starting point of the geometry in the physical model the form could be manipulated, by Rice and Stella in turn, in a way that would rationalise it, and therefore make it buildable. This is analogous to the way of working adopted for Les Nuages at La Défense. In both cases the architect wanted a flowing form and Rice's task was to find a way of transposing the initial idea into

The third element of Rice's tactics on the Groningen project that is worth noting is that he deliberately drafted in young engineers to work on the scheme with him. In typical style he believed that being young engineers they would be more flexible and more accommodating in discussions with Stella. They would, in short, not interfere with Stella's creativity.

Marc Held

The glass '*passerelle*' (bridge) at Lintas in Paris is difficult to find. It traverses a completely enclosed courtyard at high level passing over a blackened glass dome that encloses the lower levels of the courtyard. So, although it is located close to the banks

137

of the Seine, it goes unnoticed to most of the many passers by. This is unfortunate as it is an interesting little gem of a structure, built in 1986, that the new visitor to the enclosing buildings would be surprised to come across.

In terms of artistic intention this is a very different sculptural statement to that of Kurokawa's Japan Bridge that Rice was to work on five years later. Both bridges try to capture the idea of energy and dynamism, but the approaches taken to reach the end goal are very different. The Japan Bridge is generated by flowing curved lines. In contrast Marc Held works with a geometry defined by straight lines and edges colliding at different angles in three-dimensional space. In the Japan Bridge the reference is to the lanes of high-speed traffic that pass under the bridge. In Marc Held's bridge it is the client, an advertising agency, and the dynamic image that they wished to convey that act as the source for the inspiration.

Held wanted a light and fragile quality in his structure. He also needed the connection to the existing stone face of the internal courtyard to be as minimal as possible. To him the structure needed to be seen as a new intervention that floated in the space between the existing buildings: as a modern and almost ephemeral counterpoint to its stolid neighbour.

The solution devised with Rice's guiding hand was to make the top face of the

Photo: Ove Arup and Partners

Fig. 11.5. Lintas Bridge with architect Mark Held.

rectilinear glass tube do the main structural work. The top face actually looks like a cross-braced truss on its side so it is rather shallow in terms of the spanning job that it has to do. To make it work, without the depth becoming excessive, it is suspended from a prestressed steel support system that uses a pair of triangulated masts set in some distance from each end. The glass tube is then slung from the underside of the roof beam, as a set of bays that coincide with the bays in the roof structure. As a result the glass wall structures have minimal tension hangers interrupting the view out and the floor can be

made as a set of light beams spanning between each of the hangers. Designed this way, the whole structure applies loads to the supporting building that are effectively vertical, meaning that no accommodation for lateral thrusts needed to be worried about, and this, in turn, meant that interference with the existing structure was minimised.

For Rice it was the nature of the structure rather than the image of the structure that was important, and finding a solution that used materials and forms in a way that would augment the architectural idea was the number one issue to be resolved.

The lightness and dynamism in the underslung structure in the Lintas Bridge is accomplished through responding to the clear architectural view that the heavier structural gymnastics should be pushed up, and out of the way. What the pedestrian crossing the bridge senses most strongly is the lightness and the linear geometry; in other words what the architect wanted.

Kisho Kurokawa

At La Défense in Paris Peter Rice had been involved in a range of projects; part of La Grande Arche, Les Nuages and the new canopy over the CNIT entrance. In 1991 he became involved in what was to be the last of his projects in the district around La Défense that he had played a major part in shaping. That project was a bridge with the Japanese architect, Kisho Kurokawa.

This pedestrian bridge, in Valmy to the northwest of the La Défense site, is 100 m long and traverses seven lanes of traffic some 15 metres below. It joins the two parts of the financial and business district that the road bisects and runs from the Tour Pacifique building, which Kurokawa also designed, on one side, to another building on the other. The buildings on each side were capable of providing only minimal support and lateral thrusts were not possible.

The bridge was intended to be reticent of Japanese bridge design; hence its very direct name, the Japan Bridge. Kurokawa's wish was to produce a clean and gently curved form that drew inspiration from the minimal traditional small Japanese footbridge in which the walkway follows the curved structural form. He wanted a calm control and gently flowing forms to be dominant. The Japan Tower, at one end of the bridge makes strong reference to La Grande Arche next door. Kurokawa called them *'two variations on one theme: one is cartesian, the other Japanese'*. The gateway to the bridge (Fig. 11.6b) is inspired by the 'Chu Mon', the symbolic entrance to the room where the tea ceremony takes place.

The structural form adopted for the bridge brings together two tied arches. The arches are made from 900 mm deep fabricated triangular hollow steel sections for the arch and 200 mm solid bars for the tendons that tie the arch just below walkway level and prevent outward thrust. The smooth surface that the solid bars provided was felt to be more appropriate than cables given the basis of Kurokawa's inspiration. The shape of the arch and the tendons were made parabolic to make them respond as efficiently as possible to the actual dead loading. But in structural terms the problem that arose was wind loading. The walkway was to be covered by a curved glass tunnel that presented a large surface area to the prevailing winds; and the problem was

Photo: André Brown

Fig. 11.6. Japan Bridge: Tête Défense, Paris.

Chapter Eleven
Bridges and sculptures: artists and architects

Fig. 11.6a. Japan Bridge: the sweeping form.

Fig. 11.6b. Japan Bridge meets the Japan Tower through a deep, slotted, symbolic entrance.

Fig. 11.6c.
The cast-demi-nodes used to connect the struts and cables.

exacerbated by the fact that the location, at high level between two groups of multi-storey buildings, induced a kind of wind tunnel effect.

To resist the resulting problems of torsional instability under this lateral wind load meant that the structure needed some careful manipulation and modification. Rice was able to convince Kurokawa of the need for changes to make the bridge torsionally stiff, even though this would introduce elements that might detract from the clean, flowing simplicity that was wanted. The solution was to lean the two main arch ribs together so that they met at the apex, in the centre of the span, and therefore braced each other. The tendons were then heavily braced beneath the walkway, so that the resulting braced plane resisted the potential lateral instability.

The main bowstring tendons are tied to the main arches by a series of tension hangers. The walkway is connected by struts to the lower end of these hangers, so that the self-weight of the walkway pretensions the hangers giving a stable plane between the arch and the tendon.

The result of the careful manipulation of the structure and the choice of materials has ensured that, in difficult conditions, the original architectural aspirations are met. By placing the braced plane below the walkway the required torsional stiffness has been achieved with minimum visual intrusion. Added to that the clean simplicity belies the fact that the support and loading conditions were very testing. Because the support conditions were asymmetrical the component lengths and angles do vary throughout the structure even though a uniform symmetry is the overall impression. This has been achieved through the application of advanced computational techniques that allowed the analysis to be run through and modified, and then the complex geometry of the whole, and its component parts, to be defined.

One final note on materials. It would have been unlike Peter Rice to let a project slip by without the materials from which the object is made receiving due attention. Two components can be picked out here which show that due attention was accorded. First, the main steel tendons that tie the arch. The smoothness made them fit with Kurokawa's wishes, but 200 mm is a large diameter and their correct performance as principal tie members was critical. Consequently Rice specified an unusually high notch toughness to avoid failure due to brittle fracture.

The second component worthy of comment occurs at a very critical junction where the struts supporting the walkway and the hangers from the main arch meet the main bowstring tie. Rice originally conceived these, as in critical junctions in other projects, as cast steel nodes. But the contractor, Viry, suggested forging the nodes and making them in two halves which could be threaded on to the main tendon and then bolted together to clamp all of the elements meeting at the node in a sandwich. Rice could see the logic and effectiveness of this solution so was happy to adopt it. As ever if it was a good idea he did listen; no matter who the person was, or the stage of the contract.

Chapter Twelve
Stone and tension

Chapter Twelve
Stone and tension

To some, Peter Rice's experimentation with stone as a structural material is seen as strange, even bizarre, but examination of this body of work reveals that this group of structures simply represents a natural extension of his other work, and of his particular interests. In fact, by looking at these structures it could be said that we learn more about Peter Rice as an engineer and as a designer than we do by looking at much of his other output. By understanding how these structures work the student of structural design is given a lesson in how the theory can be manipulated to produce a new kind of practical solution.

In these stone structures all of the ingredients that make these buildings essentially Rice were there. It is not that Rice was the sole engineer on this project. Alistair Lenczner, Tristram Carfrae and Bruce Danzigger were, for instance, three other engineers at Arups taking a significant role in the 'Pavilion of the Future' in Seville, playing a part in a project that they, too, clearly shared a dedication to. But Rice's passions drove the direction of the work, he was the engineering fountainhead.

Rice's interest in materials was wide ranging, but can be summarised briefly as exploring the potential in how they work and how they appear. By working with materials like stone in new ways the engineer is forced to think about problems from the position of first principles. When the engineer is backed up into this corner invention is the way out, and for Rice invention meant interest, breaking the constraints of convention, and a chance for the engineer to add a new dimension to the design of built forms.

The appearance of materials also gives the designer the opportunity to extend the palette of surface texture, to reveal craft and care, and thereby to make building relate better to the human. To Rice it was imperative that we get away from the dead hand of the standard industrial process and standard solution. Using stone was yet another way to bring the designer face-to-face with this issue: a way of making the engineers think about what they wanted to achieve at a most fundamental level.

To Peter Rice the tactile and visual qualities were there to be worked with, exploited and set against other materials in a composed way. In the Seville Expo building the stone chosen was Rosa Porina granite, its name giving the hint to its pink hue, that was enhanced by flame texturing the faces that were to be exposed. At Lille Cathedral the west front was completed by a tall arch in local blue stone inside which sits an enormous 'window' made from sheets of marble cut so thin that they allow a soft, diffuse light to penetrate into the main cathedral space. At both Seville and Lille it is the stone that adds a particular visual quality, both in the appearance of the material itself, and in the new, explicit ways that the material sits within the framework of a contemporary structural solution alongside newer materials. The complementary and arresting nature of the total solution is a prime force in contributing to the architectural quality of the scheme.

Instability, or at least apparent instability, was another of Rice's loves that could be explored in the stone structures. There was, he believed, interesting mileage in taking stone, a material traditionally associated with heavy, compressive structural forms, and replacing the compressive force traditionally generated by dead weight with prestressing forces generated by fine steel structures. Our preconceptions, built on past experience, are that stone structures work in certain defined ways and employ particular forms such as the solid arch and the buttress. New materials and new methods of analysis could be employed to disturb the comfortable equilibrium that is associated with these antique forms.

In the same way that Les Serres at La Villette would perform superficially baffling structural gymnastics to hold a thin wall of glass in place, or the tension assisted structure for the Fleetguard Centre at

Quimper would be perceived as hanging onto stability by a thread, these stone structures would challenge our conventional perceptions of means of stability. Together steel in tension and stone in compression would be pared down and finely tuned to demonstrate a command of structural integrity that breathes new life into an old material.

In sum the stone structures that Rice worked on reflect his philosophy of experimentation and pushing at the boundaries, without which we limit our horizons; and without which the engineer puts unnecessary constraints on the architect's invention. The results are dramatic and thought provoking. They are innovative structures in their own right, but to Rice they were merely staging posts in a ceaseless expedition to new structural territories.

In 1988 Rice worked with Ian Ritchie on a competition for a new building at Magdalene College, Oxford. This appears to be one of Rice's earliest explorations of the potential of stone and slender tension elements working together, in this case to produce a kind of contemporary gothic semi-vault. The traditional stone arched structures in the Oxford Colleges were to be one of the design generators, but contemporary methods of analysis and load application would give a new life to these ancient forms. Late in the day the proposals were rejected in favour of an alternative design, but the ideas were to be the source of inspiration for later structures like Chiesa Padre Pio, Lille Cathedral, and the Pavilion of the Future.

Pavilion of the Future: Seville

The filigree stone and tensile steel facade that Rice and his collaborators devised for the 1992 Expo in Seville really tested the engineers' confidence in their ability to make an unconventional structural approach work. Once the delicate form in stone had been chosen in principle many problems and questions arose that challenged the engineers, and in particular Peter Rice, to have the utmost belief in their ability to find a way to turn ideas into reality.

In other projects, like the gerberettes at Beaubourg, or the tension nets at La Villette, it was not only about devising solutions to difficult, and novel problems, but it was also about convincing the authorities that the unconventional approach would work. The first of these matters was addressed through a mixture of supreme competence and resourcefulness, the second through a more direct route; Peter Rice's undoubted gift of the gab.

It was King Juan Carlos who had the idea for the Seville Expo as a celebration of the fifth centenary of Columbus' discovery of America, marking the occasion and the point

Photo: Ove Arup and Partners/Bruce Danzigger

Fig. 12.1. Seville: the Pavilion of the Future.

Chapter Twelve
Stone and tension

of his departure with an exhibition site and over a hundred pavilions from countries worldwide. In keeping with the cause for the celebration, the theme for the Expo was that of discovery.

Design of the major building at the Expo, the Pavilion of the Future (Pabellon del Futuro), was handed to the Spanish architectural firm of Martorell, Bohigas and Mackay: MBM. In their concept for the Pavilion MBM saw a curving, wave-form roof structure that would extend over the north and south pavilion halls and the central plaza between them. Along the eastern edge of the Pavilion they wanted a dramatic façade that would both announce the Pavilion, by facing towards the old city, and act as a backdrop to the expansive ornamental gardens on the site. Rice and his team were given *carte blanche* to devise a façade that would do the structural job of supporting one end of the wave-form roof structure but, more importantly, would be fitting to the Expo theme of discovery. If any engineer could match Columbus' boldness and dedication to exploring new areas it was Peter Rice.

Rice, with Ian Ritchie, had actually been in one of the other two shortlisted teams for the Pavilion competition. When MBM were selected from the three shortlisted they asked Peter Rice to join them to help create a memorable structure.

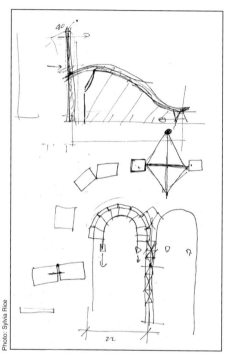

Fig. 12.2. Rice's sketches of how the Seville structure could work.

Photo: Sylvia Rice

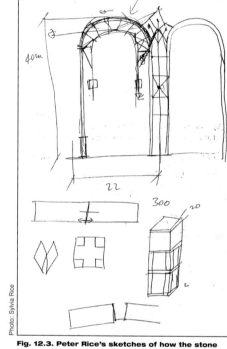

Fig. 12.3. Peter Rice's sketches of how the stone columns could be fabricated.

Photo: Sylvia Rice

For Seville the budget was tight, and the challenge was to make the façade imposing and evocative within a financial framework that many others would have found very limiting. But not Peter Rice who found his inspiration for the solution adopted in the unfinished Ajuda Palace in Lisbon. There a long wall in stone, with large arched openings, stood firmly as a free-standing element, clearly detached from any other structure that might provide support. This kind of occurrence is not unique: numerous unfinished or damaged ecclesiastical buildings can been seen around Europe with similarly imposing perforated, free-standing stone walls.

The challenge, then, was to replicate the symbolic nature of such walls but with

Fig. 12.4. Stone arch springings prior to being hoisted.

Fig. 12.5. Stone composite column being lifted into position.

new twists. It was to be about apparent contradictions; a modern ruin, a light structure made from stone.

But to start on a pragmatic level, one way that costs would be kept down would be by using small quantities of stone, and this was accomplished in two ways. First, the computational analysis would be manipulated to reduce the structure down so

that the full compressive strength of the stone would be exploited. There was no room to allow for the considerable reserves of compressive strength found in most old stone structures. Second, a relatively small proportion of the compressive load that is required to counter potentially fatal tension stresses would be provided by the traditional means of the dead weight of stone high in

the wall. Much of the compressive load would come, instead, from a prestressed tension structure connected to the stonework.

To constrain costs further, Rice decided to limit the basic stone building blocks, taking a standard 20 cm x 20 cm cross-section unit and an 80 cm x 80 cm solid slab as the basis for developing an open format column and arch structure. Because the structure was being pared down to the minimum it was essential that the jointing worked evenly and efficiently. Some joints would need rigid connections which would be provided by modern epoxy resins, and other joints would be debonded (with short stainless steel locating dowels to resist shear) to prevent tension developing in them.

So, two new materials, epoxy resins and stainless steel, would be employed, in typical Rice fashion, to help make an old material function in an unconventional way. But in order for the structure to work in the way intended it was necessary that the stone be cut to a high degree of accuracy, and this was a matter of concern to Rice's team early in the project. They were worried that the local quarry might not be able to cut the stone elements to the small tolerances required in the thin glued joints that would be fundamental to ensuring the integrity of the facade structure. In the end their fears were unfounded, and the stonemasons,

147

Chapter Twelve
Stone and tension

Arquiedra, were to produce the stone pieces reliably to the tolerances specified. The unexpected problem was that testing of the stone revealed that there was a plane of natural weakness that could lead to premature fracturing under load. In most cases stone is used as a cladding material in contemporary buildings, so the strength properties are relatively unimportant. By returning stone to its role as a load-carrying material Rice was exposing an Achilles heel in this particular stone that would otherwise be unimportant. But he would see this as a consequence of his approach, and simply another matter to be tackled by engineering thinking; an extra aspect that generated a unique and personal attachment of the engineer to the engineered product.

The solution was to be found in taking care with regard to the direction of loading applied to each stone piece. This was a pragmatic solution to a pragmatic problem. On a more general level, the approach that was developed to devise appropriate structural forms shows a level of inventiveness and adaptation which is particularly impressive.

The structure
The façade is composed of 11 semi-circular stone arches, with a radius of 8·66 metres. These arches span onto pairs of 28 m high stone columns, which are spaced at 22·4 m centres, each arch springing from the inner of the two columns in the adjacent pairs. Together the columns and the arches form a flat vertical plane: a kind of stage set. To give stability to the flat plane, the pairs of columns are linked by horizontal steel tubular cross-bars which are, in turn, connected to a triangular latticework of thin steel elements that act as a vertical trussed

Fig. 12.7. Centre section of the stone arch being hoisted into position.

Fig. 12.6. During construction with the wave-form trusses temporarily propped.

Fig. 12.8. Centre section of the arch in position.

cantilever, extending up from ground level to the head of the column. In this regard the steel and stone structures act as a composite tower.

On top of the outside pair of columns partial arches begin to spring, but are truncated well before they meet in the centre of the arch. These partial arches are simply a reference to the incomplete or demolished building that is the project's generator. Although they serve this role they have two other important structural functions which they satisfy quite cleverly. First, the additional weight at the head of the column adds to the prestress in the column that is needed to overcome any potential tensile stresses which stone is poorly equipped to deal with. Second, they allow the steel truss to extend up past the head of the columns, giving depth to the structure behind the plane of the arch. This provides the arch with the means of resisting the potential problem of out-of-plane instability.

The structural function of the arches was to carry the ends of the wave-form trusses that support the roof over the areas behind the façade. These are not insignificant loads. For the arch ring to be only 80 cm deep for a span of 17·3 m the load thrust line has to follow the semi-circular shape of the arch almost exactly. If the thrust line falls outside the inside face (*intrados*) or the outside face (*extrados*) then a hinge develops

which can lead to collapse. This means that the loads applied to the arch have to be applied at regular intervals, and applied evenly to the arch.

In order to turn the point loads, generated by the ends of the wave-form trusses, into a load that is evenly distributed around the arch, the ends of the trusses are held up by cables that are suspended from the inside face of the stone arch. These cables are, in turn, suspended from a set of short steel bars, which together form a ring that is concentric with the primary arch ring. The ring of steel bars is held in position by a set of radial tie rods, spaced at 15° intervals, which are fixed to the primary arch through slots cut in the inner face of the stone work. This carefully managed loading arrangement means that, as far as possible, the load path is close to the centre line of the arch and, additionally, the loads applied to the supporting columns are vertical.

The weight of the wave-form trusses and their cladding was insufficient to overcome the potential wind uplift forces so ballast weights were added to the ends of the trusses to counteract this possible uplift. This meant that not only is the stone arch apparently light and frail, but also the roof structure and these additional weights are held aloft by the tension rods that extend upwards to the tension ring. They are, most deliberately, hanging by the finest of threads.

Different loading conditions mean that perfect symmetry cannot always be guaranteed. One way in which this is addressed is by putting an additional tie in between the end of the tension arch ring, diagonally downwards to the supporting column. This braces the structure against out-of-balance loads in any pair of adjacent wave-form trusses.

The structural acrobatics being performed here are what gave the façade structure its arresting power. These acrobatics are needed to maintain

Fig. 12.9. Arch and tie system before tensioning of the tied inner ring.

Photo: Ove Arup and Partners/Bruce Danziger

Chapter Twelve
Stone and tension

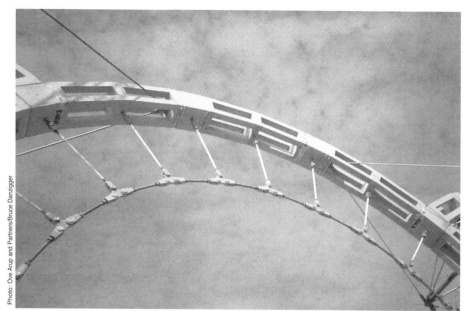

Photo: Ove Arup and Partners/Bruce Danzigger

Fig. 12.10. The tension ring completed.

two other details are worthy of particular mention, for different reasons.

The tie rods in the trussed steel structure meet the steel strut in the assembly at steel node points. The node connector had radial symmetry about the strut but, in addition, the node needed to allow the ties to meet at the junction at a variety of spatial angles. It also needed to allow for tensioning of the rods, and to allow for minor angular imperfections. The standard solution in this kind of situation is to use the U-shaped shackle with a pin across the legs of the shackle. This, of course, was not the solution that Rice went for. Instead, a special steel connector was designed which allowed tension to be adjusted in the rods; it accommodated angular imperfection and

compressive forces within such a relatively fine and delicate stone structure and this means that the engineers had to devise particular and finely tuned solutions. They did this with conviction, and without compromise.

As in most Rice buildings it is the detail that gets particular attention. All around the façade structure interesting details present themselves. The slotted connections joining the radial steel rods to the primary arch and the elegant pinned connections that connect the radial rods to the inner tension ring complement the whole structure and add to its richness. But

Fig. 12.11. The structure against the Seville skyline.

Photo: Ove Arup and Partners/Bruce Danzigger

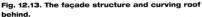

Fig. 12.12. The slender façade and ends of the wave-form trusses.

Fig. 12.13. The façade structure and curving roof behind.

actually cost less than the standard solution. It also gave a joint which was more satisfying visually, smoother and cleaner than its standard counterpart.

Also hidden in the joint were special conical spring washers which applied the correct pretensioning force when they were fully compressed. But, in addition, as they reached full compression they behaved like solid washers, giving the overall structural system the total stiffness that it was designed to achieve.

Final notes

The apparent fragility and unconventional nature of the filigree façade was an essential part of its dramatic impact. But as in other projects, despite Rice's track record, the authorities needed some convincing that the idea really could work.

Various weapons were required in the battle to convince. One of these was computational methods. The engineering team used the FABLON Dynamic Relaxation program to analyse the structure, but had to incorporate new types of elements that had edges which ran diagonally across the arch. These special elements only worked in compression, so that if tension was developed on one face of the arch one pair of adjacent elements effectively disappeared. This left the other element in the adjacent pair to make a connection over a small area

on the compression edge, hence producing the hinge condition that was required. With the computer program suitably modified to take account of the particular behaviour of stone structures it was possible for the Arup engineers to run a wide variety of load cases to prove the robustness of the structure under a wide range of potential loading conditions.

Having been just about convinced that the delicate structure could actually stand, the authorities were still not totally convinced that it could be constructed. To allay these fears the Rice team developed a system of prefabricating the column and arch in a series of sub-units. These sub-units had to be hoisted into position, and to do this the team designed cradles to support the units while they were lifted. These were needed because the structure was, indeed, very fragile. It was designed within very tight constraints, and only worked in the designed way when the total structural system had been composed; when all the stone, steel and tension rods had been connected and pretensioned.

But, even after all of these reassuring measures had been taken by Rice's team the Expo chiefs in Seville remained cautious. They insisted that the ends of the wave-form roof should be propped, and not hung from the arches, until the very last minute. Of course, when the props were removed the

(left margin, rotated) Photo: Ove Arup and Partners/Bruce Danziger

(left margin, rotated) Photo: Ove Arup and Partners/Bruce Danziger

151

Chapter Twelve
Stone and tension

Photo: Ove Arup and Partners/Steve Abernethy

Fig. 12.14. The structure gets to perform its gymnastics routine.

structure worked. The chiefs should have spoken to the SOCOTEC sceptics who questioned the feasibility of the cast steel gerberettes at Beaubourg. After all if you wanted a conservative structure you did not ask Peter Rice.

Cathédrale Notre Dame de la Treille, Lille

In the original construction of Lille Cathedral, the west face had never been completed. But it was self evident that the completion of that face was going to be a core element when the scheme for the refurbishment of the cathedral square was announced. The architect for the cathedral square scheme was Pierre Louis Carlier, and he called in Peter Rice's team at RFR to devise a way of providing a new face to the square to replace the ugly infill that stood there.

By this time RFR had a well-established reputation in France. They were seen as a practice that could turn architectural concepts that were sculptural, unconventional and dramatic into real objects. At Lille what Carlier needed was a west façade that would be read as cool and reverential on the inside of the cathedral, but bold and sculptural on the outside. A solution that would make the cathedral work as a cathedral, but would also act as a foil for the urban composition of the eponymous square on to which it fronted. The design team was enriched by the inclusion of an artist, Ladislas Kinjo, and by developing the ideas for Lille in conjunction with stained window craftsmen.

In many ways this project was one that brought together the kind of design team that Rice pleaded for. Engineers and architects, craftsmen and artists; all around the table, together from the start. All with particular skills and experience that could, if managed well, contribute additional quality to the final product. It was the kind of environment that would have been close to Rice's ideal.

As at Seville, the brief for Lille was one where the combination of the ancient material, stone, and the new material, steel, allowed them to be played off against each other to provide an engaging tension; one where control over material, both old and new, could be demonstrated and exploited. The final design, as enacted, has a very tall arch in local blue stone, called 'pierre bleu' that, in shape, follows the lines of the cathedral nave. Its stability in-plane is achieved through its arched form; it acts

like a very tall portal frame which is prestressed through a system of cables, that can be read as such on the outside of the cathedral, but are seen as cast shadows from the inside.

The steel bracing system of cables is actually made in stainless steel, and was designed in collaboration with the artists for the scheme. The geometry and relief created by the steel prestressing net is something that evolved through discussions with the artists, so that the form satisfied both artistic intention and structural function.

A curtain of marble, cut into sheets that are so thin that they are translucent, is suspended from the arch. Set within the marble skin, at a high level, is a large circular stained glass window; a rose window. The result of this combination within the arch is that the marble transmits a soft light into the nave, while the rose window casts a circlé of coloured light.

The idea of admitting light through thin sheets of marble into a space in this way is reminiscent of the technique used in the Taj Mahal. At Lille, though, the net of prestressed cables casts shadows on to the marble, and this is read as a pale grid when seen from the inside; a clear reminder of the fact that here old materials and new technologies are being brought together in a mutually supportive way.

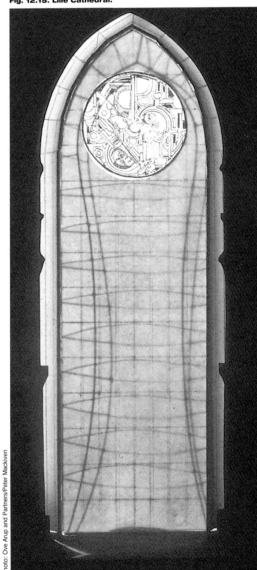

Fig. 12.15. Lille Cathedral.

Photo: Ove Arup and Partners/Peter Mackiven

153

Chapter Twelve
Stone and tension

In structural terms Lille represents an extension and amalgamation of ideas. There are echoes of La Villette in the tension based transom trusses, but this time the trusses are anchored back to a stone arch rather than steel columns. And the thin film of brittle, fragile material that they complement is translucent marble rather than glass. At the top of the arch the pure circular hole that is punched through the marble film to allow the rose window to be inserted has to be formed in some way. The technique used is reminiscent of the Seville Expo and has radiating ties running up to the intrados of the arch, effectively suspending the ring that forms the opening for the circular stained glass window.

This is yet another example of experimentation in one project being reapplied in a later one; a feature that is evident in much of Rice's work. The Seville Expo was not an isolated dead end, but an opportunity to add new materials, methods and ideas to the burgeoning armoury that Rice had to draw on. Here, at Lille, he was able to evolve the ideas of marrying pretension in steel with stone in compression and add in the experience of projects in glass and steel. The composition that arises from this is one that is sophisticated and challenging. In marrying old and new technologies

Photo: André Brown

Fig. 12.15a. New end wall to Lille Cathedral.

Photo: André Brown

Fig. 12.15b. Braced blue-stone arch with supended marble sheets.

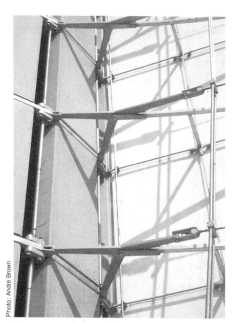

Photo: André Brown

Fig. 12.15c. Detail showing the connection of the tension net to the stonework.

Photo: André Brown

Fig. 12.15d. The suspended marble curtain and rose window from inside the cathedral.

together in a controlled and sculptural form it provides an interface between the old cathedral and the new elements in the cathedral square. It is not simply a plaything. It is both an architectural device and a remarkably imposing sculptural object in its own right.

Chiesa Padre Pio

The home of the 'miracle priest' Padre Pio, the town of San Giovanni Rotondo, set in the foothills that rise from the Apulian Plain (Tavoliere delle Puglie), has become an important pilgrimage venue. In 1991 Renzo Piano was commissioned to design a building that would be both a church and a pilgrimage centre for the town in a location close to the monastery where Padre Pio lived and the chapel where he preached.

Clearly, this was to be a building that would need to respond to a range of cultural, religious and contextual pressures. At this time Piano and Rice were both close friends and close too in their maturing approach to the spectrum of design influences which Piano described as putting together...

'the tesserae in the mosaic of the profession, balancing out professionalism, experimentation, history and form...'[68]

The Padre Pio project was about creating a congregating place that could both act as a focus for the collection of many hundreds of people in a sermon on the mount kind of scenario, and provide a symbolic focus for religious dedication. The term 'church' belies its massive scale and its function. The walls of the church, like those at Assisi, were to be a physical landmark with which pilgrims could make actual and spiritual contact.

The design sketches and initial thoughts that Piano had for the building show a spiral form in plan. The open end of the spiral spilled out at the front of the church on to a gently sloping stone piazza where large outdoor congregations would collect. Inside the church the spiralling forms become tighter and interweave, coming to a final focus at the altar; the sacred heart of the church.

The structure was seen by Piano as a series of intersecting, radiating arches that would define ribbed vaults in the enclosed spaces and reinforce the spatial concepts. The conventional structural solution would have been a series of reinforced concrete arches but this would not have appealed to Rice and Piano. Other projects revealed their shared interests and concerns that were now explicit. Bringing together new materials and new forms with old materials was one of these concerns. Other concerns included articulation of the composing elements,

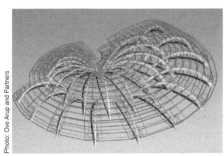

Photo: Ove Arup and Partners

Fig. 12.16. Intersecting arches at Chiesa Padre Pio.

the marks of craft in the articulated pieces (referred to by Rice as *'trace de la main'*) and using contemporary analytical methods to develop new structural forms. All of these would be there to work with in the Padre Pio project.

For Rice the structural solution had more to do with the Menil Gallery structure than the stone and steel structures at Seville and Lille. The arches at Padre Pio were conceived as curved backbones, made (in the final form) as 4·2 x 1·8 m stone segments that acted in composite action with tensioned steel ties and timber struts, spanning approximately 50 m. The triangulated shape in cross-section has, in the way the materials are

Chapter Twelve
Stone and tension

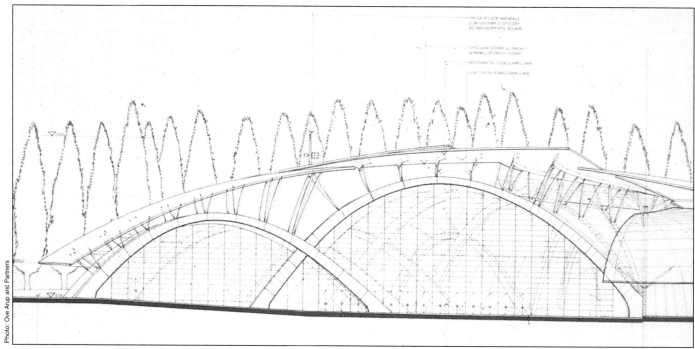

Fig. 12.17. Elevational drawing of Padre Pio.

Photo: Ove Arup and Partners

composed, a strong parallel in the Menil roof beams. At Menil the lower, compression boom was made in ferrocement; in Padre Pio the segmented stone elements would take on a similar role. The ductile iron trussed pieces on the upper part of the Menil beams would be replaced by steel and timber; the tension elements taking the form of thin steel ties, and the compressive elements as timber shaped elements; their larger cross-section reflecting the fact that they were doing a different job to that of the

steel, a job that required them to resist buckling.

The composite nature of the arch ribs set up the potential for both articulation and for playing the historic game. The historic monastery and chapel were to be joined on the site by this new partner, the Padre Pio church. But within the church that mixture of contexts was to be reflected in the collaboration of stone with steel and timber; new forms and materials set against the traditional. The depth and articulation in the structural form that was conceived also gave the opportunity to work with light inside the building. As at Menil the quality and nature of natural daylight, and the way that it penetrated the enclosed space, would add a further dimension to the design.

During the design work on Chiesa Padre Pio Peter Rice became ill and was not able to see the project through to its final design stages. But Renzo Piano makes it clear that the final form of the project is almost entirely as Rice conceived it. As such it represents another significant landmark in Rice's collaboration with architects and in pushing the boundaries of design through the exploration and exploitation of materials.

Fig. 12.18. The articulated ribs of the church.

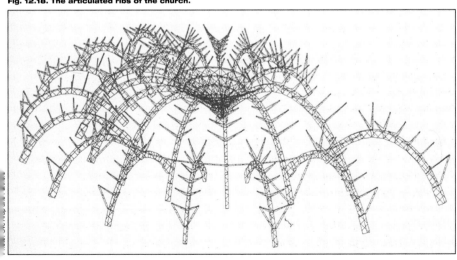

Chapter Thirteen
The legacy

Chapter Thirteen
The legacy

An archetypal Rice building

If Peter Rice had been asked to choose a favourite building it would probably have been one in which he had a particular personal interest. In that category would figure the Moon Theatre, which he developed a particular fondness for, or the Beaubourg building that saw him come of age as a designer–engineer. Both of these projects also saw him develop long-term friendships which would also have made them special; with Humbert Camerlo, the Director of the Full Moon Theatre, and with Renzo Piano and Richard Rogers at Beaubourg.

But there is for me one building which captures, in a single instance, many of the key qualities that Peter Rice strove for. A building which perhaps more than any other reflects the guiding principles and driving passions behind the hand and mind of Peter Rice. In terms of defining the legacy that Peter Rice passed on to others it is a particularly good example.

It is not the Sydney Opera House. At the Opera House he played a key role on site, getting the building constructed, and solving a range of practical and computational problems to realise the high quality that is still evident in the building today. But he was learning from those around him, as well as giving.

Nor was it Centre Pompidou. Again, this was a building where others made significant design decisions, and it was a building that was a means of developing and establishing ideas. It was in some ways the start of a process of collaboration with a group of notable architects in which Rice acted as enabler and foil, providing the technical and pragmatic skills that were provided in such a way that they nurtured the evolution of a maturing architectural philosophy.

The building, or more precisely buildings, that I believe encapsulate many of the definitive qualities that testify to the influence of Peter Rice are the glasshouses at the Parc André Citroen in Paris. The park, also known as Parc des Cevennes, was built on the site of the old Citroen car plant and was completed in 1993. Set in an urban regeneration area, and surrounded by offices and housing, it is a place that is popular with Parisians. It is a calm and tranquil area composed of hard and soft landscape, still and flowing water, and built and natural forms.

The architect, Patrick Berger, worked closely with the RFR team. Together they responded to the different landscape themes in the park, designing a family of eight greenhouses. The family is split into two types. There are two larger greenhouses set on raking stone platforms with flowing

Photo: André Brown

Fig. 13.1. Small glass houses at Parc André Citroen.

Photo: André Brown

Fig. 13.2. Major glasshouses.

sheets of water and fountains to compliment the setting. Then there are a further six smaller, 9 m high, cuboid structures dotted around. These are simple but striking forms; a single central column supporting a grid of cantilevered beams from which the thin glass skin is draped.

The tension structures that support the glass walls are clearly part of the family of structures that have Rice parentage, the lineage of which can be traced back to the walls at Les Grandes Serres at La Villette. The structures are not radically different to those that precede them. Thin tension cables are pushed apart by minimal spreaders. The whole approach here is not radical but

Fig. 13.3. Glasshouses with stone plinth and fountains in the foreground.

Photo: André Brown

Chapter Thirteen
The legacy

evolutionary. The anchorages are adapted to suit the local conditions. Positive and negative wind pressures, along with lateral instability, are dealt with in familiar ways. Familiar, that is, to those who know the family history. For those who do not there is the immediate visual tension as these thin sheets of brittle glass are seen to be supported by this fine net of tension wires that are pushed apart by minimal spreaders. To add to the drama the glass is balanced on the fingertips that extend out from the net.

In the roof the tension structure has echoes of the earlier Louvre roof. In the corners of the buildings are four columns. Stout affairs, these. Vertical cantilevers

Fig. 13.4. Transom trusses.

Fig. 13.5. Short transom trusses and connection to the plane of glazing.

Photo: André Brown

Fig. 13.6. Mullion trusses, transom trusses and column.

planted like telegraph poles in the ground, their rich, shiny and slightly undulating timber surface a significant contrast to the refined minimal technology in the thin cable structures that sit between them.

Then, even to the technologist, there is the problem of working out how these damned tension structures work. If it is the first time that the visitor has seen one of them there is the process of going through the logic step by step in the mind.

'If the wind load is positive and pushing the glass face inwards then this cable goes into tension...but when the wind causes suction that other cable is in tension...Oh yes, and that

Photo: André Brown

Fig. 13.8. Anchorage detail at floor level.

Fig. 13.9. Roof structure at Parc André Citroen.

Photo: André Brown

Fig. 13.7. Junction of the trusses and connection to the glazing.

Chapter Thirteen
The legacy

element stops the whole of the structure flipping out of plane (failing due to lateral instability)...'

At least that is how it is for me when I try to follow the logic of how it all works.

Then you have to convince the Building Authorities that it really does work: that you cannot use a simple linear analysis because certain members disappear under certain loading due to a thin cable going into compression. So the structure changes its form and a relatively complex kind of analysis is needed to describe the behaviour. A combination of mathematical ability and a smooth tongue are needed to convince the sceptic. Enter Peter Rice, starring in his own little film.

Now, this kind of debate about how and whether these tension structures work took place principally around the time of La Villette. Peter Rice does not claim a particular significance or sole role in the Parc André Citroen buildings. These are not the important issues. What matters is that they bring together a group of principles that Peter Rice espoused. They reveal the kind of rich architecture that can result from adopting the Rice-enhanced approach.

Rice liked to introduce a new material or use a new material in a different way whenever the project allowed it. This was part of his crusade to get craft back into what he saw as on over-industrialised

process. In the André Citroen glasshouses this aspect is not taken on as radically as it is in the use of stone in the Seville arch or ductile iron at Menil. But the wooden surface on the columns has a human crafted quality that was so appealing and important to Peter Rice.

But, having said that, it would be easy to forget that Rice was acutely aware of his professional responsibilities and the pragmatic requirements that go hand-in-hand with technological development. The façades at Parc André Citroen were, because they were considered so innovatory, subject to a rigorous ATEX (Avis Technique d'Experimentation) testing programme. And the glazing system used is one that had recently been patented by RFR.

In the end what is important is that the glasshouses are buildings of quality. They are products that testify to the benefits of engineer–architect collaboration in the Rice style.

Fig. 13.10. Star shaped connector at the corner of the panes of glazing.

Photo: André Brown

Luxembourg Museum

In 1990, towards the end of Peter Rice's active involvement in projects at RFR he worked on a major project in Luxembourg. This was at the time when the Roissy Charles de Gaulle Airport project was on site and nearing completion. The Luxembourg scheme was to provide an opportunity to continue the developments which were evident in the evolving design approach at RFR.

The attachment of the glazing system to the steel frame with tight tolerances and minimal support structure, which was being evolved through schemes like Roissy Charles de Gaulle Airport, were to be taken on here. And the technique for restraining uplift used in schemes like the glasshouses at Parc André Citroen was to find another variant here too.

Fig. 13.11. Luxembourg Museum of Modern Art.

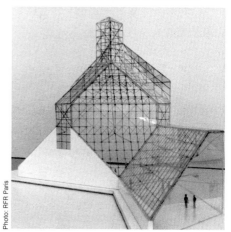

Photo: RFR Paris

The initial work on the scheme which took place in early 1990 was for a glazed atrium roof for the modern Art Museum, a steel and glass structure commissioned by the Ministry of Pubic Works. Design work on the scheme had been progressed to a reasonably advanced stage but had then been put on hold. In mid 1998 the scheme got the go ahead. So, six years after his death, a scheme that shows clear evidence of the deft mind of Peter Rice was resuscitated and will serve as a reminder of the kind of quality and excitement that his input brought to a building design.

Other projects currently active at RFR, like the Luxembourg Museum project, show the continued influence of Peter Rice. The practice continues the evolutionary process that he began; the experimentation and seeking to rework and enhance ideas relating to structure and materials continues. In this regard two projects are worth referring to.

The first is the design for the new Strasbourg Museum. As with the Luxembourg Museum this was a project that Rice worked on the early stages of, but was not able to see through to completion.

The Pyramide Inversée at the Louvre was one line of evolution from La Villette. There the walls were drawn in at angles and the tension was provided by self-weight. In the same way the Strasbourg Museum Nef takes the glass and steel structure at La Villette as a starting point. But added to that is the rather testing problem for this kind of structure, that the building is located in a potential earthquake region. A major change, which Rice had to implement to address this issue, was to use double, rather than single glazing. To Rice this was not perceived as a

Fig. 13.12. Strasbourg Museum.

Photo: RFR Paris

Fig. 13.13. Luxembourg Airport.

Photo: RFR Paris

problem, but simply an interesting extension of engineering design ideas; another little prod at the boundaries of what engineering and architecture could, together, achieve.

The second project of this kind that is worthy of a little attention is the new terminal building for Luxembourg Airport. This is notable, not because of Rice's direct involvement, although he was involved in the early stages of the design, but because of his indirect influence. It is interesting to compare the roof structure here (Fig. 13.13) with the roof structure at the de Menil Gallery (Fig. 7.15). The curved 'leaves' of ferrocement in the de Menil Gallery serve two functions; they are part of the roof structure and they deflect direct sunlight so that a good quality of diffuse light is radiated into the gallery. For Luxembourg, Rice's colleagues and prodigies have taken the idea further and devised a set of curved elements, made as a steel latticework rather than as composite iron–ferrocement structures. But the principals are there; the curved form gives the structure depth, so that is can act as a roof beam/truss as well as a light deflector.

Rice would have been pleased to see ideas like that for the Luxembourg Airport building emerging out of RFR. The atmosphere there remains one that is experimental and challenging. The motivations that drove Peter Rice continue to drive RFR.

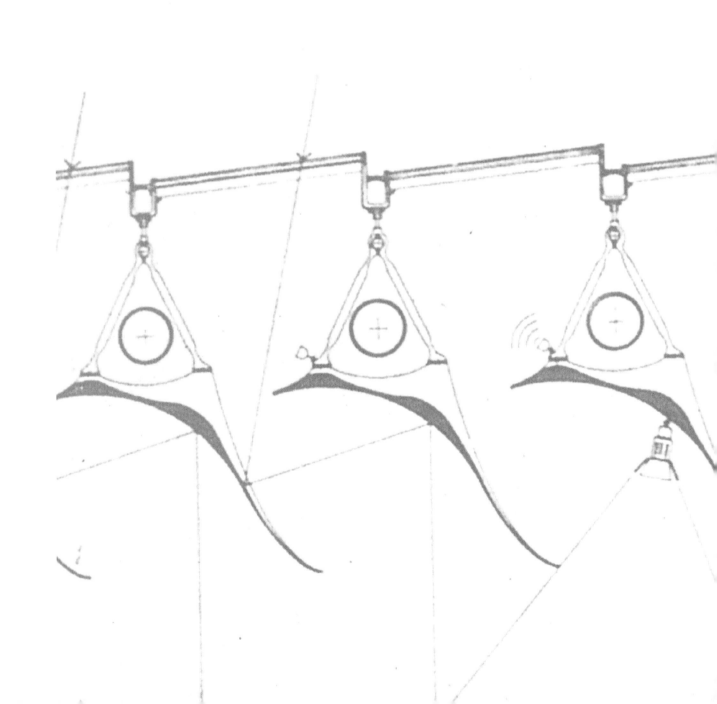

Chapter Fourteen
Conclusions

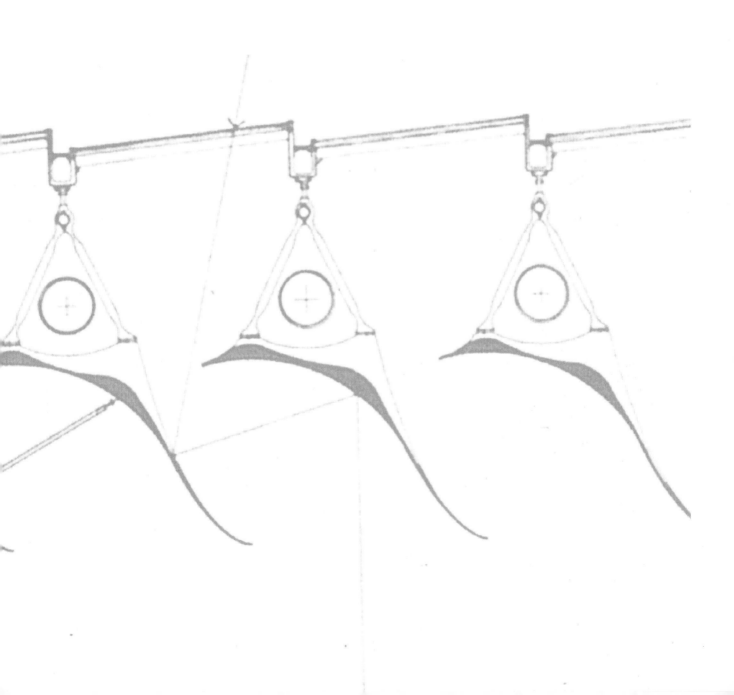

Chapter Fourteen
Conclusions

How can we describe Peter Rice's contribution to contemporary architecture? In refining and evolving the ideas embodied in High Technology it was undoubtedly central. One of the arms of the modern tradition, the rationalist approach as espoused by Le Corbusier, was a victim of its own edicts, unable to break free and evolve from the constraining legacy of prescribed modern functionalism. Conversely, the High Technologists were able to move on and take on new ideas. It needed an engineer like Peter Rice to enable this evolution to happen. He was able to bridge the gap between mathematical complexity and design aspiration, between the engineering pragmatism and architectural ideal.

Despite his strong associations with France and French architecture we are fortunate that Peter Rice saw fit to ignore the views of France's most famous architect, Le Corbusier. In 1925 Le Corbusier said '*an engineer should stay fixed and remain as a calculator, for his particular justification is to remain within the confines of pure reason*'. It would be difficult to hold a view that contradicted Peter Rice's beliefs more extremely than this one. For Rice pure reason meant uninspiring. The tension structure supporting the Bioclimatic Façades at La Villette are typical of Rice; as are, though perhaps less so, the gerberettes at the

Pompidou Centre. Apparent contradiction, intrigue and mild deception were all qualities that Rice regarded as entertaining. They were the spices that gave the meal its distinctive flavour. They were ways of avoiding mindless and unchallenging repetition.

It is not that being a calculator worried Peter Rice. The opposite would be closer to the truth. He was described by Tristram Carfrae, who worked with him as engineer on this and other projects, as an accomplished and confident mathematician. Going back to first principles to calculate the stone arch for the Seville Expo was part of the challenge, and therefore part of the design. Disentangling the complex computer analysis of the tension nets that were to hold the tents (Les Nuages) in place at La Défense: these were not simply challenges to Rice, they were yet another source of design modification and design decision making. For Corbusier calculation was the end of the engineering production line. For Rice it was another tool in the armoury that he could turn to for guidance or to play tricks with.

Other engineers would find refuge in the tried and tested, and what was known, Peter Rice would deliberately seek the unknown. New materials, new analytical techniques and unconventional projects would be the basis for his projects if he could make it so. Confident of his abilities to find a way back, he was most satisfied wandering

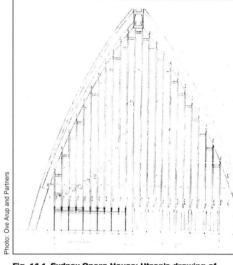

Photo: Ove Arup and Partners

Fig. 14.1. Sydney Opera House: Utzon's drawing of the glass end wall.

off into the engineering unknown with the architect's concept as the light that guided him.

His diverse interests outside engineering were the source of his uniqueness. Inspiration could come from horse racing or art, maths or poetry. While other engineers might have regarded such passions as a luxury, they were the defining qualities that made the approach that Peter Rice took so inventive. He could draw on a wide range of sources for inspiration, or as prototypes. An example of that is the spider's webs research that he got involved with.

Sir Jack Zunz had had a long working relationship with Peter Rice at Arups, guiding aspects of the early Rice career. In 1989 Zunz's daughter Marion wondered if Peter Rice might like to talk to Dr F. Vollrath, a zoologist that he knew at the University of Oxford. Vollrath's interest was in spider's webs. Peter Rice got extremely enthusiastic about the work. To him this was an opportunity to learn more about a structure that had gone through 180 million years of evolution. It was a chance to look at how these nets absorbed kinetic energy and at the potential architectural forms that these nets might take. So he got to work and with his assistant, Lorraine Lin, he undertook a process of modelling and analysis which Vollrath noted made an important contribution to zoological science.

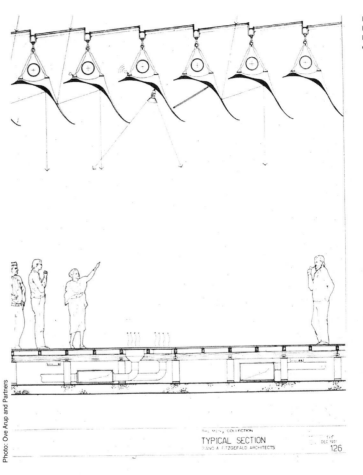

**Fig. 14.2.
Menil Collection:
Piano's drawing of the
critical cross-section.**

Photo: Ove Arup and Partners

Chapter Fourteen
Conclusions

So, while Jack Zunz was '*getting more and more cautionary about the costs in manpower and computer time*',[69] Rice was planting design seeds that could germinate in later projects. His prodigy, Tristram Carfrae, said that it was typical of his approach; to work out a solution today and wait for the problem to arise tomorrow.

Years later RFR were given part of the design work for the new national sports stadium in Paris, the Stade de France. One aspect of the work was to provide barriers on the façade of the high level external walkways that ran around the perimeter of the stadium. Now the conventional solution might have been a rigid steel or concrete frame supporting some kind of secondary structure. But the RFR solution was simple and elegant. A web of energy absorbing steel mesh was fitted in a single layer. This allowed views out and meant that the walkways were airy, but best of all it embodied a neat solution to a major problem; that of crowds of spectators getting crushed in the case of an emergency escape. In other notable disasters rigid barriers have led to injury and death. Here the barrier was a web that absorbed energy and flexed if crowds pushed against it. A giant spider's web was being used to soak up the potential impact, and the research undertaken years before was not wasted.

Peter Rice cared about his profession. His compassion and passion to produce great engineering are evidenced both in his work and in what his collaborators say about him.

In Peter Rice's very particular case the answer to the question 'What was his

Fig. 14.3. Model of the clouds: tension and fabric structures on the Parvis at La Défense.

contribution to contemporary architecture?'
is that he did not simply contribute. He was
'*a key figure, who along with Norman Foster,
Richard Rogers, Michael Hopkins, Ian Ritchie
and a few others... have given rise to a new
British architecture.*' He did this from the
position of engineer and specialist. He has
planted seeds in both fields. We need those
seeds to germinate. We need more engineers
who think about their profession in the way
that Rice did, and to continue to hold out
the hand of friendship and support to the
creative architect.

Richard Rogers opens his web site with
a page on the practice philosophy, and
concludes that page with a quote. That
quote is by Peter Rice and it is apt to end
this book with the same quote.

'*Good teams are made up of different
people, people whose separateness and attitude
complement each other and who, by their
individual willingness to work together and
accept the presence and contribution of all the
others, for a while at least make possible real
momentum.*'

Endnotes

Preface

1. **Rice P.** (1992). Speech at 1992 RIBA Gold Medal Presentation. *RIBA Journal*, Sept.

2. **Rice P.** (1986). *Exploring the boundaries of design*. Pidgeon Audio Visual 15/8610, London.

3. **Sommer D., Stocher H.** and **Lutz W.** (1994). *Ove Arup and Partners: Engineering the built environment*. Birkhåuser Verlag, Basel, Switzerland.

4. **Zunz J.** (1993). Peter Rice. *Arup Focus*, 3 (92), Autumn Supplement.

Chapter One

5. *Exploring materials: the work of Peter Rice, Royal Gold Medallist 1992*. Catalogue to accompany exhibition, prepared by B-A. Campbell and J. MeMinn with the Ove Arup Partnership.

6. **Zunz J.** (1993). Peter Rice. *Arup Focus*, 3 (92), Autumn Supplement.

7. **Rice P.** (1989). A celebration of the life and work of Ove Arup. RSA Journal, June, 415–437, reprinted in *Arup Journal*, **25**, 1, Spring 1990, 43~47.

8. **Rice P.** (1989). A celebration of the life and work of Ove Arup. *RSA Journal*, June, 415–437, reprinted in *Arup Journal*, **25**, 1, Spring 1990, 43~47.

9. **Tassel A. A.** and **Rice P.** (1985). Master of the Hi-tech style: Britain's Peter Rice. *SA Construction World*, Apr., 54–61.

10. **Various** (1992). Peter Rice: tributes to a great structural engineer. *Architects' Journal*, 4 Nov.

11. **Rice P.** *The Structural Engineer*. Lecture given on 17 Jan. 1990 as part of the 'Engineers in Design' series.

12. **Rice P.** (1992). Speech at 1992 RIBA Gold Medal Presentation. *RIBA Journal*, Sept.

13. (1987). Il Punto di Vista di Peter Rice (An engineer's view). *L'Arca*, 5, Apr., 70–75.

14. (1991). Bridging the gap: Rethinking the relationship of architect and engineer. *Building Arts Forum*, Van Nostrand Reinhold, New York.

Chapter Two

15. (1987). Soaring aspirations: Peter Rice talks about his admiration for the Gothic masons. *Royal Academy Magazine*, 17, Winter, 29–30.

16. (1985). The nature of materials: interview with Peter Rice. *Architecture d'Aujourd'hui*, 237, Feb., 10–17.

17. **Rice P.** (1989). Building as craft, building as industry. *Guggenheim Museum Proc.* 87–89 (New York).

18. **Rice P.** (1986). *Exploring the boundaries of design*. Pidgeon Audio Visual 15/8610, London.

19. **Buchanan P.** (1983). Patscenter, Princeton. *Architectural Review*, **174**, 1037, 43–47.

20. **Rice P.** (1992). Speech at 1992 RIBA Gold Medal Presentation. *RIBA Journal*, Sept.

21. **Carfrae T.** (1999). Personal communication.

22. **Wernick J.** (1999). Personal communication.

23. **Rice P.** (1993). The Iago approach. *Domus*, 749, May, 17–24.

24. **Day A., Haslett T., Carfrae T.** and **Rice P.** (1986). Buckling and non-linear behaviour of space frames. *International Conf. on Lightweight Structures in Architecture*, Sydney, 1, Aug., (LSA 86), 775–782.

25. **Dunster D.** (1997). *Arups on engineering*. Ernst and Sohn, Berlin.

26. **Rice P.** (1990). Unstable structures. *Columbia Documents of Architecture and Theory*, **1**, Rizolli, New York, 71–90.

27. **Rice P.** (1993). The Iago approach. *Domus*, 749, May, 17–24.

Chapter Three

28. **Messent D.** (1997). *Opera House: Act One*. David Messent Photography, Sydney.

29. **Rice P.** (1994). *The engineer imagines*. Artemis Press, London.

30. **Rice P.** *The Structural Engineer*. Video of lecture given on 17 Jan. 1990 as part of the 'Engineers in Design' series.

31. **Rice P.** (1986). *Exploring the boundaries of design*. Pidgeon Audio Visual 15/8610, London.

32. **Nutt J.** (1992). Rice and the Sydney Opera House. Obituary: *Sydney Times*,

Chapter Four

33. **Rice P.** (1989). Building as craft, building as industry. *Guggenheim Museum Proc.* 87–89 (New York).

34. **Rice P.** (1989). Building as craft, building as industry. *Guggenheim Museum Proc.* 87–89 (New York).

35. **Piano R., MacCormac R.** and **Rogers R.** (1992). Royal Gold Medal Address. *RIBA Journal*, **99**, Sept., 26–29.

36. **Happold E. F.** and **Rice P.** (1973). Introduction. *Arup Journal*, Centre Beaubourg Special Issue, **8**, 7, June, 2–3.

37. **Ahm P. B., Clarke F. G., Grut E. L.** and **Rice P.** (1979, 1980). Design and construction of the Centre National d'Art et de Culture Georges Pompidou. *Proc. Instn. Civ. Engrs*, Part 1, 66, 557–593 (Nov. 1979), Part 1, 68, 499–505 (Aug. 1980).

38. **Rice P.** (1986). *Exploring the boundaries of design.* Pidgeon Audio Visual 15/8610, London.

39. **Rice P.** (1994). *The engineer imagines.* Artemis Press, London.

40. (1985). The nature of materials: interview with Peter Rice. *Architecture d'Aujourd'hui*, 237, Feb., 10–17.

41. **Rice P.** (1994). *The engineer imagines.* Artemis Press, London.

Chapter Five

42. **Rice P.** (1994). *The engineer imagines.* Artemis Press, London.

43. **Ferrier J.** and **Lavalou A.** (1990). Figures Brittaniques: Peter Rice. *Architecture d'Aujourd'hui*, 267, Feb., 114–121.

44. **Sarfarti A.** (1998). Keynote speech at the *16th eCAADe Conf.*, Paris, Val de Marne, Sept.

45. **Rice P.** *The Structural Engineer.* Lecture given on 17 Jan. 1990 as part of the 'Engineers in Design' series.

46. *Exploring materials: the work of Peter Rice, Royal Gold Medallist 1992.* Catalogue to accompany exhibition, prepared by B-A. Campbell and J. MeMinn with the Ove Arup Partnership.

Chapter Six

47. (1986). La Villette: Cité des Sciences et de Industrie. *Techniques et Architecture*, 364, Feb–Mar., 60–129.

48. **Rice P.** and **Dutton H.** (1990). *Le Verre Structurel.* Editions du Moniteur, Paris.

49. **Wernick J.** (1999). Personal communication.

Chapter Seven

50. **Rice P., Barker T., Guthrie A.** and **Noble N.** (1983). The Menil Collection, Houston Texas. *Arup Journal*, **18**, 1, 2–7.

51. **Rice P., Lenczner A.** and **Carfrae T.** (1991). The San Nicola Stadium, Bari. *Steel Construction Today*, July, 57–160.

Chapter Eight

52. **Rice P.** (1994). *The engineer imagines.* Artemis Press, London.

53. **Young J.** and **Thornton J.** (1984). Design for better assembly: (5) Case Study: Rogers and Arups. *Architects' Journal*, **180**, 36, 5 Sept., 87–94.

54. **Rice P.** and **Thornton J. A.** (1986). Lloyd's redevelopment. *Structural Engineer*, **64A**, 101, Oct., 265–281.

55. **Rice P.** (1994). *The engineer imagines.* Artemis Press, London.

56. **Rice P.** (1990). Equilibre et Tension, (Calatrava at Stadelhofen, Zurich). *Archithese*, 2–90, 84–96. Also published in German as Gleichgewicht und Spannung, but not in English.

57. **Rice P.** (1990). Equilibre et Tension, (Calatrava at Stadelhofen, Zurich). *Archithese*, 2–90, 84–96. Also published in German as Gleichgewicht und Spannung, but not in English.

Chapter Nine

58. **Davey P.** (1987). Operation Overlord's. *Architectural Review*, **82**, 1087, Sept., 40–49.

59. (1987). Stratégie de l'Araignée. *Architecture d'Aujourd'hui*, 252, Sept., 78–79.

Chapter Ten

60. **Rice P.** (1986). *Exploring the boundaries of design.* Pidgeon Audio Visual 15/8610, London.

61. **Nutt J.** (1992). Rice and the Sydney Opera House. Obituary: *Sydney Times*,

62. **Rice P.** and **Dutton H.** (1990). *Le Verre Structurel.* Editions du Moniteur, Paris.

63. **Spring, M.** Transport de Light, *Building*, 15 July 1994, pp. 42-45.

Endnotes

Chapter Eleven

64. **Rice P.** (1986). *Exploring the boundaries of design.* Pidgeon Audio Visual 15/8610, London.

65. **Rice P.** (1994). *The engineer imagines.* Artemis Press, London.

66. **Rice P.** (1994). *The engineer imagines.* Artemis Press, London.

67. **Rice P.** (1986). *Exploring the boundaries of design.* Pidgeon Audio Visual 15/8610, London.

Chapter Twelve

68. **Piano R.** and **Lampugnani V. M.** (1995). *Renzo Piano: 1987–1994.* Birkhåuser, Basel.

Chapter Fourteen

69. **Rice P.** (1986). *Exploring the boundaries of design.* Pidgeon Audio Visual 15/8610, London.

Selected List of Works

1957
Sydney Opera House, Sydney, Australia
Architect: Jorn Utzon, Hall Todd & Littlemore, Hanson & Todd Pty Ltd, New South Wales Government, C. P. Wedderburn, Rudder Littlemore & Rudder Pty Ltd
Consulting Engineers: Ove Arup & Partners

1967
Crucible Theatre, Sheffield
Architect: Renton Howard Wood Associates
Consulting Engineers: Ove Arup & Partners

1969
Amberley Road Children's Home, London
Architect: Renton Howard Wood Associates
Consulting Engineers: Ove Arup & Partners

1970
National Sports Centre, Crystal Palace, London. Proposals for a new stadium roof structure
Architect: Greater London Council
Consulting Engineers: Ove Arup & Partners

Circus 70, Victoria Embankment, London
Architect: Casson Conder & Partners.
Consulting Engineers: Ove Arup & Partners

New Arts Centre, Warwick University, Coventry
Architect: Renton Howard Wood Levin Partnership
Consulting Engineers: Ove Arup & Partners

Perspex Spiral Staircase, Jeweller's Shop, Jermyn Street, London
Architect: Godfrey H. & George P. Grima
Consulting Engineers: Ove Arup & Partners

Super Grimentz Ski Village, Valais, Switzerland A new ski village for 5000 visitors, including a skating rink, swimming pool and car park
Architect: Godfrey H. & George P. Grima
Consulting Engineers: Ove Arup & Partners

1971
Conference Centre, Mecca
Architect: Rolf Gutbrod Architects
Consulting Engineers: Ove Arup & Partners

Special structures advice to Frei Otto and others on pneumatic and cable structures including 'The City in the Arctic'
Architect: Frei Otto
Consulting Engineers: Ove Arup & Partners

Centre Pompidou, Paris
Architect: Piano & Rogers
Consulting Engineers: Ove Arup & Partners

St Katharine Dock, London
Architect: Renton Howard Wood Levin Partnership
Consulting Engineers: Ove Arup & Partners

1972
World Trade Centre, London Conversion of
St Katharine Dock House
Architect: Renton Howard Wood Associates
Consulting Engineers: Ove Arup & Partners

1976
Johannesburg Jumbo Jet Hangar, South Africa
Consulting Engineers: Ove Arup & Partners

1977
Fruili Housing, Italy
Architect: Renzo Piano Building Workshop
Consulting Engineers: Ove Arup & Partners

Pilkington study. Study to carry out prototype development of roofing units using glass fibre reinforced cement
Architect: Richard Rogers & Partners
Consulting Engineers: Ove Arup & Partners

1978
Hammersmith Interchange, London
Architect: Foster Associates
Consulting Engineers: Ove Arup & Partners

Lloyds of London Redevelopment, City of London
Architect: Richard Rogers & Partners
Consulting Engineers: Ove Arup & Partners

Industrialised construction system for Vibrocemento, Perugia.
Piano & Rice

11 Rigo Quartier, Perugia, Italy
Piano & Rice

Fiat VSS Experimental Vehicle, Turin, Italy
Piano & Rice

Fleetguard, Quimper, France
Architect: Richard Rogers & Partners
Consulting Engineers: Ove Arup & Partners

Patscentre, Princeton, NJ, USA
Architect: Richard Rogers & Partners
Consulting Engineers: Ove Arup & Partners

1979
Victoria Circus Shopping Centre, Southend-on-Sea, Essex
Architect: Alan Stanton
Consulting Engineers: Ove Arup & Partners

Educational Television Programme,
RIA Television: The Open Site
Piano & Rice

An experiment in Urban Reconstruction for UNESCO, Otranto
Piano & Rice

1980
Design for Burano Island, Venice
Piano & Rice

Fabric Roof Canopy, Schlumberger
Headquarters, Montrouge, France
Architect: Renzo Piano Atelier de Paris
Consulting Engineers: Ove Arup & Partners

1981
Serres and Toiture Accueil de la CM des Sciences et Industrie, La Villette
Architect: Adrien Fainsilber
Consulting Engineers: RFR and Ove Arup & Partners

IBM Travelling Pavilion
Architect: Renzo Piano Building Workshop
Consulting Engineers: Ove Arup & Partners

Stansted Airport Terminal Building
Architect: Foster Associates
Consulting Engineers: Ove Arup & Partners

Menil Collection Museum, Houston, Texas
Architect: Renzo Piano Building Workshop
Consulting Engineers: Ove Arup & Partners

1982
Alexander Pavilion, London
Architect: Terry Farrell
Consulting Engineers: Ove Arup & Partners
Peter Rice Consulting Engineer

1983
Alton Towers, Alton, Staffordshire Jet Star 2 Building
Architect: Griffin Jones Associates
Consulting Engineers: Ove Arup & Partners

Clifton Nurseries Roof, Covent Garden, London
Architect: Terry Farrell
Consulting Engineers: Ove Arup & Partners

Pavilions, Tate Gallery, London
Architect: Alan Stanton
Consulting Engineers: Ove Arup & Partners

1984
122 St John Street, London
Architect: Eva Jiricna
Consulting Engineers: Ove Arup & Partners

Environment and Motorway, Berlin, West Germany. Feasibility
study for motorway acoustic protection system & solar heating
for adjacent properties
Architect: Pascal Sh6nning
Consulting Engineers: Ove Arup & Partners

Ballsports Stadium, Berlin, Germany
Architect: Christoph Langhof, Architekten
Consulting Engineers: Ove Arup & Partners

1985
Emplacement, North Queensferry, Lothian, Scotland
Architect: Ian Ritchie Architects
Consulting Engineers: Ove Arup & Partners

Louvre, Paris. Design of steel structure to carry a glass roof over
courtyards
Architect: I. M. Pei with Michel Macary

Lord's Mound Stand, London
Architect: Michael Hopkins & Partners
Consulting Engineers: Ove Arup & Partners

Aztec West Reception Building, Bristol
Architect: Michael Hopkins Architects
Consulting Engineers: Ove Arup & Partners

Ray Square, Narrow Street, London
Architect: Ian Ritchie Architects
Consulting Engineers: Ove Arup & Partners

Atrium Roof, Conflans, Saint Honoré, France
Architect: Valode et Pistre
Consulting Engineers: Ove Arup & Partners

1986
Fabric Canopy, St Louis/Basle, France
Architect: Aéroport de Paris/Paul Andreu

Tête Défense, La Défense, Paris, Nuage Leger
Architect: J. O. Spreckelsen and Aéroport de
                         Paris/Paul Andreu
Consulting Engineers: Ove Arup & Partners and RFR

Nuage Parvis, Paris, La Défense
Architect: J. O. Spreckelsen and Aéroport de
                         Paris/Paul Andreu
Consulting Engineers: RFR and Ove Arup & Partners

Passerelle Lintas, Paris
Architect: Marc Held
Consulting Engineers: RFR

Galeries du Parc de La Villette, Paris
Architect: Bernard Tschumi
Consulting Engineers: RFR

Central House, Whitechapel High Street, London
Architect: Ian Ritchie Architects
Consulting Engineers: Ove Arup & Partners

Football Stadium, Bari, Italy
Architect: Renzo Piano Building Workshop
Consulting Engineers: Ove Arup & Partners

IBM 'Ladybird' Travelling Exhibition, Italy
Architect: Renzo Piano Building Workshop
Consulting Engineers: Ove Arup & Partners

Apartment for John Young, London
Architect: John Young
Consulting Engineers: Ove Arup & Partners

Opera, Bastille, Paris. Studies for acoustic ceiling
Architect: Carlos Ott
Consulting Engineers: RFR

Usine Centre, Epone, France. Steel warehouse hypermarket
structure
Architect: Richard Rogers & Partners
Consulting Engineers: Ove Arup & Partners

Usine Centre, Nantes, Loire Atlantique, France. Steel warehouse
hypermarket structure
Architect: Richard Rogers & Partners
Consulting Engineers: Ove Arup & Partners

Floating Restaurant, Jubilee Gardens, London
Architect: Richard Rogers & Partners
Consulting Engineers: Ove Arup & Partners

European Synchrotron Radiation Facility, Grenoble, France
Architect: Renzo Piano Atelier de Paris
Consulting Engineers: Ove Arup & Partners

1987
Parc Citroen Cevennes, Greenhouse, Paris
Architect: Patrick Berger
Consulting Engineers: RFR

Passerelles Front de Seine, Paris
Design and Consulting Engineers: RFR

Couverture du Chateau de Falaise, Normandy
Architect: Decaris
Consulting Engineers: RFR

Selected List of Works

Verriere, Musée des Beaux Arts de Clermont-Ferrand
Architect: A. Fainsilber and Gaillard
Consulting Engineers: RFR

Salle Polyvalente, Nancy, France. Design for 70 metre span
cable-braced roof
Architect: Foster Associates
Consulting Engineers: Ove Arup & Partners

58 metre Motor Yacht. Computer aided design work for the
stability of the yacht
Naval Architects: Martin Francis
Consulting Engineers: Ove Arup & Partners

Ravenna Sports Hall, Ravenna, Italy
Architect: Renzo Piano Building Workshop
Consulting Engineers: Ove Arup & Partners

Competition for Aircraft Hangars, Abu Dhabi, United Arab
Emirates
Architects: Aéroport de Paris/Paul Andreu
Consulting Engineers: RFR and Ove Arup & Partners

Azabu & Tomigoya Structure, Tokyo
Architect: Zaha Hadid
Consulting Engineers: Ove Arup & Partners

Office/apartment Block, Lecco, Italy
Architect: Renzo Piano Building Workshop
Consulting Engineers: Ove Arup & Partners

G.R.C., Mossy, Essone, France
Architect: Richard Rogers & Partners
Consulting Engineers: Ove Arup & Partners

UNESCO Laboratory/Building Workshop, Genoa, Italy
Architect: Renzo Piano Building Workshop
Consulting Engineers: Ove Arup & Partners

Atrium, Offices for Bull Corporation, Avenue Gambetta, Paris
Architect: Valode et Pistre
Consulting Engineers: RFR and Ove Arup & Partners

1988
La Grande Nef, Tête Défense, Paris
Architect: Jean-Pierre Buffi
Consulting Engineers: RFR

TGV/RER Charles de Gaulle, Roissy
Architect: Aéroport de Paris/Paul Andreu
Consulting Engineers: RFR and Ove Arup & Partners

Tours de Liberté, Paris
Architect: Hennin et Normier
Consulting Engineers: RFR and Ove Arup & Partners

Façade de la B.P.O.A. Rennes
Architect: O. Deccl + B. Cornette
Consulting Engineers: RFR

Franconville project, France
Architect: Cuno Brullman, d'Architect and Arnaud
                     Fougeras Lavergnolie Architects
Consulting Engineers: Ove Arup & Partners

Chur, Switzerland. Glazed roof canopy, bus/rail station
Architect: Robert Obrist and Richard Brosi
Consulting Engineers: Ove Arup & Partners and RFR

Piazza, Stag Place Site B, London
Architect: Richard Rogers Partnership
Consulting Engineers: Ove Arup & Partners

Sistiona, Italy. Studies for conversion of disused quarry & sea
frontages into resort
Architect: Renzo Piano Building Workshop
Consulting Engineers: Ove Arup & Partners

Crown Princess Liner, Italy. Design of the superstructure for the
refurbishment of a liner
Architect: Renzo Piano Building Workshop
Consulting Engineers: Ove Arup & Partners

Usine Centre Herblain, France
Architect: Richard Rogers & Partners
Consulting Engineers: Ove Arup & Partners

Museum of Contemporary Art, Bordeaux, France
Architect: Valode et Pistre
Consulting Engineers: Ove Arup & Partners

Centre Commercial, Pontoise, Val d'Oise, France
Architect: Richard Rogers Partnership
Consulting Engineers: Ove Arup & Partners

Pearl of Dubai, Dubai, United Arab Emirates
Architect: Ian Ritchie Architects
Consulting Engineers: Ove Arup & Partners

Harbour Museum, Newport Beach, California, USA
Architect: Renzo Piano Building Workshop
Consulting Engineers: Ove Arup & Partners

RAF Northolt, London New Air Terminal
Architect: Property Services Agency
Consulting Engineers: Ove Arup & Partners

Kansai International Airport Terminal, Japan
Architect: Renzo Piano Building Workshop
Consulting Engineers: Ove Arup & Partners

Europe House, World Trade Centre, London
Architect: Richard Rogers Partnership
Consulting Engineers: Ove Arup & Partners

Marseilles Airport Terminal Building, France
Architect: Richard Rogers Partnership
Consulting Engineers: Ove Arup & Partners

Grand Louvre, Paris Observation Pavilions
Architect: Michael Dowd
Consulting Engineers: Ove Arup & Partners

National Portrait Gallery, 39–45 Orange Street, London
Architect: Stanton Williams
Consulting Engineers: Ove Arup & Partners

Bercy 2, France, Roof system for Shopping Centre
Architect: Renzo Piano Atelier de Paris
Consulting Engineers: Ove Arup & Partners

Studies for Discoveryland Mountain, Eurodisney
Architect: Disney Imaginering
Consulting Engineers: Ove Arup & Partners and RFR

Competition for Patinoire d'Albertville, France. Covered skating
rink for Winter Olympics
Architect: Adrien Fainsilber
Consulting Engineers: RFR and Ove Arup & Partners

Moon Theatre, Provence, France. Designer and Theatre
Director: Humbert Camerlo
Consulting Engineers: Ove Arup & Partners and RFR

Hull Design, Staquote British Defender Yacht, International off-shore racing, Whitbread Round World Race
Naval Architects: Martin Francis
Consulting Engineers: Ove Arup & Partners

1989
Auvent, Ville de Verclun
Architect: Pierre Colboc
Consulting Engineers: RFR

Verriere Inclinée (sloping glazed façade and roof), Shell, Ruell Malmaison
Architect: Valode et Pistre
Consulting Engineers: RFR

Passerelle Est/Ouest, Galleries Nord/Sud and Bridge, Parc de La Villette, Paris
Architect: Bernard Tschumi
Consulting Engineers: RFR

L'Oreal Factory, Aulnay, France
Architect: Valode et Pistre
Consulting Engineers: Ove Arup & Partners

Pavilion of the Future, Expo '92, Seville, Spain
Architect: Martorell Bohigas Mackay
Consulting Engineers: Ove Arup & Partners

Industrial Buildings, Tomaya, Japan
Architect: Sugimur
Consulting Engineers: Ove Arup & Partners

Queen's Stand, Epsom Racecourse, Surrey
Architect: Richard Harden Associates
Consulting Engineers: Ove Arup & Partners

Toronto Opera House, Canada
Architect: Moshe Safdie & Associates
Consulting Engineers: Ove Arup & Partners

BIGO Tent, Colombo 500, Genoa, Italy
Architect: Renzo Piano Building Workshop
Consulting Engineers: Ove Arup & Partners

Pharmacy, Boves, France
Architect: Ian Ritchie Architects
Consulting Engineers: Ove Arup & Partners

Helios VII. Feasibility study for summer house based on solar sculpture
Consulting Engineers: Ove Arup & Partners

Spiders' Webs Research Project
Zoologist: Dr Fritz Vollrath
Consulting Engineers: Ove Arup & Partners

Office Building, Stockley Park, London
Architect: Ian Ritchie Architects
Consulting Engineers: Ove Arup & Partners

Kurfurstendamm, Berlin, Germany, Façade study
Architect: Zaha Hadid, Stefan Schroth
Consulting Engineers: Ove Arup & Partners and RFR

Mitsubishi Tokyo Forum, Competition
Architect: Richard Rogers Partnership Ltd
Consulting Engineers: Ove Arup & Partners

New East Gallery, Natural History Museum, London
Architect: Ian Ritchie Architects
Consulting Engineers: Ove Arup & Partners

Villette Serre Study, Serres de La Villette, Paris
Architect: Kathryn Gustafson
Consulting Engineers: Ove Arup & Partners

Utsurohi, La Défense, Paris, France. Sculpture
Artist: Iyo Miyawaki
Consulting Engineers: Ove Arup & Partners

1990
Competition for Station Square, Oberhausen, Germany
Design Engineering: RFR

Auvent et Façades du CNIT, Paris
Consulting Engineers: RFR

Campanile, Place d'Italie, Paris. Tower structure with sculpture
Architect: Kenzo Tange, Macary, Menu
Artist/Sculptor: Thierry Vide
Consulting Engineers: Ove Arup & Partners

Atrium Roof, Grand Ecran, Place d'Italie, Paris
Architect: Kenzo Tange
Consulting Engineers: RFR

IMAX Cinema and Leisure Building, Liverpool
Architect: Richard Rogers Partnership
Consulting Engineers: Ove Arup & Partners

Groningen Museum, Groningen, Holland. Preliminary study of roof structure
Architect: Alessandro Mendini
Artist: Frank Stella
Consulting Engineers: Ove Arup & Partners

Padre Pio Chiesa, St Giovanni Rotondo, Puglia, Italy
Architect: Renzo Piano Building Workshop
Consulting Engineers: Ove Arup & Partners

Centre Culturel de la Pierre Plantee Bibliotheque, Vitrolles, France
Architect: Ian Ritchie Architects
Consulting Engineers: Ove Arup & Partners

Mobile Sculpture, Genoa 500
Artist: Sasumo Shing
Consulting Engineers: Ove Arup & Partners

Development of glass system with Asahi Glass, Japan
Design: RFR and Ove Arup & Partners

1991
Aéroport Charles de Gaulle, Terminal 3, Roissy, France
Architect: Aéroport de Paris/Paul Andreu
Consulting Engineers: RFR

Atrium, Centre D'Art Contemporain, Luxembourg
Architect: I. M. Pei
Consulting Engineers: RFR

Lamp post, Esch sur Aizette, Luxembourg
Architect and Town Planner: Prof. Sievents
Design and engineering: RFR

Pyramide Inversée, Grand Louvre, Paris. Inverted glass pyramid sculpture suspended over underground public circulation area at museum
Architect: I. M. Pei
Consulting Engineers: RFR

Selected List of Works

Japan Bridge, Paris
Architect: Kisho Kurokawa
Consulting Engineers: RFR and Ove Arup & Partners

Piscine, Levallois Perret
Architect: Cuno Brullman
Consulting Engineers: RFR

Rompe R4/T4, Parc de La Villette
Architect: Bernard Tschumi
Consulting Engineers: RFR

Passerelle, Mantes-la-Jolie
Architect: Michel Macary
Consulting Engineers: RFR

Ligne Meteor, Paris
Architect: Bernard Kohn
Consulting Engineers: RFR

Aérogare de Luxembourg
Architect: Paczowski
Consulting Engineers: RFR

Hotel Module d'Echanges, Roissy, France
Architect: Aéroport de Paris and Paul Andreu
Consulting Engineers: RFR

Cathédrale Notre Dame de la Treille, Lille, France
Architect: Pierre-Louis Carlier and Artist L. Kinjo
Consulting Engineers: RFR

Bus Fluviaux, Lyon
Architect: Sernaly
Consulting Engineers: RFR

Technocentre Renault, Guyancourt, France
Architect: Valode et Pistre
Engineering: Ove Arup & Partners and RFR

Lille TGV Station, Roof design
Architect: SNCF Jean-Marie Duthilleul
Consulting Engineers: Ove Arup & Partners and RFR

Demountable Pavilion, Museum of the Moving Image, London SE1
Architect: Future Systems
Consulting Engineers: Ove Arup & Partners

Brau & Brunnen Tower, Berlin, Germany
Architect: Richard Rogers & Partners
Consulting Engineers: Ove Arup & Partners

Atrium glazing, 50 Avenue Montaigne, Paris, France
Architect: O. Vidal
Consulting Engineers: RFR

Atrium for new Renault Headquarters
Architect: Valode et Pistre
Consulting Engineers: RFR

Grand Louvre, Louvre Museum, Paris. Natural lighting system for new museum
Architect: Pei Cobb Freed & Partners
Consulting Engineers: Ove Arup & Partners

Ajaccio Airport
Architect: Aéroport de Paris and Paul Andreu
Consulting Engineers: RFR

Glazing for Digital Headquarters, Geneva, Switzerland
Technical assistance to SIV, Italy
Consulting Engineers: RFR

Fabric Sculpture, Sainsbury's, Plymouth, Devon
Architect: Dixon Jones
Consulting Engineers: Ove Arup & Partners

1992
Façade pour l'extension du Palais de Congres, Paris
Architect: Olivier Clement
Consulting Engineers: RFR

Passerelle, Levallois-Perret, France
Architect: Caubel
Consulting Engineers: RFR

Atrium, Museum of Modern Art, Strasbourg
Architect: Adrien Fainsilber
Consulting Engineers: RFR

Cultural Centre, Albert, France
Architect: Ian Ritchie Architects
Consulting Engineers: Ove Arup & Partners

Western Morning News, Plymouth
Architect: Nicholas Grimshaw and Partners
Consulting Engineers: Ove Arup & Partners

Bibliography

(1984). Interview with architect John Young and engineers Peter Rice and John Thornton. *Architects' Journal*, **180**, 361, 5 Sept., 87–94.

(1989). Pei as you learn: Exhibition Pavilion, The Louvre, Paris — Architects: Michael Dowd with Peter Rice. *Building Design*, 17 Feb., 28–29.

(1989). Interview with Peter Rice. *Le Moniteur*, 4453, 31 Mar., 46–47.

(1992). Conversation with Peter Rice. In Thornton, C. et al. *Exposed structures in building design*. McGraw-Hill, New York, 152-158.

Boles D. D. (1987). The point of no return. *Progressive Architecture*, July, 94–97.

Chipperfield D. (1988). Tecnologia i Arquitectura = Technology and Architecture. *Quaderns d'Arquitectura i Urbanisme*, 178, July–Sept., 12–25.

Cook P. (1987). Glass Link. *Blueprint*, 37, May, 26–27.

Cook P. (1988). Passerela in Vetro à Parigi. *Abacus*, 4, 14, June–July, 5, 10, 66–71.

Ellis C. (1987). Unused abattoir becomes a bristling science museum. *Architecture*, Sept., 85–87.

Ferrier J. (1992). Peter Rice le grand ingénieur high tech est décédé. *Architecture d'Aujourd'hui*, 284, Dec., 26–28.

Fisher T. (1989). A non-unified field theory. *Progressive Architecture*, Nov., 65–73, 135.

Gabato C. and Rice P. (1990). Bari. *Architecktur Aktuell*, **24**, 139, Oct., 82–85.

Glancey J. (1992). Peter Rice. *The Independent*, 29 Nov.

Goulet P. (1985). La Nature des Matériaux: de l'Acier Moulé an Polycarbonate — Entrentien avec Peter Rice. *Architecture d'Aujourd'hui*, 137, Feb., 10–17.

Groak S. (1984). Engineers as pioneers. *Architects' Journal*, **180**, 45, 7 Nov., 48–49.

Gruber D. (1992). Reflections on a consummate artisan. *Progressive Architecture*, Dec., 84–87.

Knobel L. (1983). Striking through the mask. *Domus*, 637, Mar., 10–23.

Le Corbusier (1930). *Precisions: on the present state of architecture and city planning*. Paris, Cres et Cie (English translation Cambridge MA MIT Press, 1991).

Papadernetriou P. C. (1987). The responsive box. *Progressive Architecture*, May, 87–97.

Pawley P. (1989). The secret life of the engineers. *Blueprint*, 55, Mar., 34–36.

Pélissier A. (1986). Peter Rice à La Villette. *Techniques et Architecture*, 364, Feb.–Mar., 134–135.

Piano R. (1987). *Renzo Piano: the process of architecture*. Exhibition Catalogue, 9H Gallery, London, 16 Jan.–15 Feb.

Pink J. (1992). Future shock. *Architectural Review*, **190**, 1144, June, 53–54.

Plowman A. (1985). Pick of the 'Cover' Storeys. *Glass Age*, Apr.

Relph-Knight L. (1984). Paris, Texas…. *Building Design*, 26 Oct., 18–19.

Rice P. (1971). Notes on the design of cable roofs. *Arup Journal*, 6, 4, 6–-10

Rice P. (1977). Materials in use: Fire protection and maintenance at the Centre Pompidou. *RIBA Journal*, **84**, 11, Nov., 476–477.

Rice P. (1980). The structural and geometric characteristics of lightweight structures. Ove Arup Partnership: Arup Partnerships' Seminar: *Lightweight structures*, Mar.

Rice P. (1980). Lightweight structures: Introduction. *Arup Journal*, **15**, 3, Oct., 2–5.

Rice P. (1981). Long spans and soft skins. *Consulting Engineer*, 45, 7, 10–12.

Rice P. (1986). Engineers and architects working together with some examples of spatial and lightweight structures. *International Conf. on Lightweight Structures in Architecture*, Sydney, 1, Aug.

Rice P. (1987). Menil Collection museum roof: Evolving the form. *Arup Journal*, **22**, 2, Summer, 2–5.

Rice P. (1987). The controlled energy of Renzo Piano. *Renzo Piano: The process of architecture*, Exhibition Catalogue, 9H Gallery, London.

Rice P. (1990). Constructive intelligence. *Arch +*, 102, Jan., 37–51.

Rice P. (1991). Practice and Europe. *The Structural Engineer*, **69**, 23, 3 Dec., 400–401.

Rice P. (1991). Peter Rice — Great Britain. In *Architecture Now*, Arcam Pocket, (Architectura & Natura), 127–128.

Rice P. (1991). Menil Collection museum roof: Evolving the form. *Offramp*, **1**, 4, 117–119.

Rice P. (1992). Dilemma of technology. In **Tzonis A.** & **Lefaivre L.** (eds), *Architecture in Europe since 1968: Memory and Intervention*, Thames & Hudson, London, 36–41.

Rice P. and Grut L. (1975). Main structural framework of the Beaubourg Centre, Paris. *Acier Stahl Steel*, **40**, 9, Sept., 297–309.

Rice P. and Peirce R. (1976). The Barrettes of Centre Georges Pompidou. South African Instn Civ. Engrs, Fifth Quinquennial Regional Convention: *Innovations in civil engineering*, Stellenbosch University, 28–30 Sept.

Rice P., Lenczner A., Carfrae T. and Sedgwick A. (1990). The San Nicola Stadium, Bari. *Arup Journal*, **25**, 3, Autumn, 3–8.

Ritchie I. (1992). Peter Rice 1935–1992. *Building Design*, 30 Oct.

Sudjic D. (1980). High tech with a human face. *Architects' Journal*, **172**, 28, 9 July, 74–75.

Waters B. (1981). Setting the sails. *Building*, **241**, 37, 3–34.

AV and exhibition media

Rice P. (1982). *La Cité en Luminaire, La Villette*. Video, RFR.

The Late Show. BBC TV, 1992. Profile of Peter Rice on winning the RIBA Gold Medal.

Index

Index